MINAMATA
Kumamoto Gakuen University
Minamata Studies Booklet

Guidebook

A Walk to Learn about Minamata Disease

Open Research Center for Minamata Studies, Kumamoto Gakuen University

Guidebook: A Walk to Learn about Minamata Disease
by Open Research Center for Minamata Studies, Kumamoto Gakuen University
Translated by Kayoko Matsuda, Harue Sano, and Hiroko Saisho
Translation edited by Craig Armstrong

Contributors
Shigeharu Nakachi, Masanori Hanada, Masami Tajiri, Yukari Inoue, Takuji Umeda, Minoru Takagi, Atsushi Tanaka, Sadao Togashi, Hideyuki Nakamura, Yuukou Nakamura, Naoko Hamaguchi, Masazumi Harada, Takashi Miyakita, Yoshihiro Yamashita

© 2019 by Open Research Center for Minamata Studies, Kumamoto Gakuen University
First published in Japanese by Kumamoto Nichinichi Shimbun in 2019.
This English edition published by Kumamoto Nichinichi Shimbun in 2020.

Coverage Cooperation: Aiko Okamoto, Nobuhiro Fujimoto
Photograph offer: Kumamoto Nichinichi Shimbun, Iwao Onitsuka, Takeshi Shiota , Tadao Matsuzaki

The Date of Issue: 31 March 2020

ISBN 978-4-87755-604-4 C0336
Printed in Japan

Preface

It was May 1956 when two innocent girls of two years and 11 months and five years and 11 months fell ill at Tsukinoura in Minamata City, which subsequently led the official discovery of Minamata disease. This year marks the 50th anniversary.

In the 50's, Japan declared itself "no longer postwar", while rushing up the ladder of economic growth without due attention to any other considerations. Technological development kept its magnificent pace according with the economic evolution, which was, it should be acknowledged, making our lives affluent and convenient. The country was just about to depart from the phase of postwar deprivation, and approach the age of plenty. Media was also making a major shift from print and radio to television. Ownership of private cars, once a fantasy, was almost within the reach of the middle class. Vending machines made their dazzling appearance on the street.

In parallel with such economic growth, Japan concluded its peace treaty with Soviet Union, and was admitted to the United Nation. Japan was proceeding toward becoming a major economy. Falling behind this glamorous facade, however, the negative impacts were piling up in many areas of Japan. 70's saw the nationwide rise of antipollution campaigns and pollution litigation, which were simply the eruption of accumulating pressure and the uprising of the people abandoned by their government.

In Minamata, plastic and vinyl products, then state of the art and providing of more convenience, made their grand debut. At the same time, the primary industries such as fisheries were dwindling, with more population moving into the cities. Outbreak of Minamata disease, against such backdrop, had symbolic significance.

Half a century later, we are now reexamining the case of Minamata disease from various standpoints to address today's problems. Aiming for a base for research open to the participation of citizens and passing its achievement to the local communities, we have established the Open Research Center for Minamata Studies at Kumamoto Gakuen University in Kumamoto City, and the Onsite Research Center for Minamata Studies in Minamata City. These centers have been pursuing a wide range of activities, part of which is the publication of this booklet.

With the rapid progress and diffusion of IT, the modes of learning and participation have been drastically changing. In this modern world as typified by the internet, we have purposely chosen the printed media for publication, not only because we wanted to provide diverse information surrounding Minamata disease as widely as possible, but also because Minamata disease was a negative aspect of the cutting-edge technologies. We hope you find this printed material (booklet) somewhat nostalgic, friendly, easy to read, accessible, yet reflecting the present days.

We expect and readily accept the criticism that this booklet is not a polished publication edited by a research center. This lack of polish is, in fact, part of the characteristic of an open research center, and in a sense, part of the purpose. I sincerely hope that many people will feel free to read and use this booklet.

May 1, 2006
Masazumi Harada
Director of the Open Research Center for Minamata Studies, Kumamoto Gakuen University

Contents

Before Walking in Minamata

This year marks the 64th anniversary of the official announcement of the recognition of the disease that is now known as Minamata disease, and the 114th anniversary of Chisso Corporation's establishment. What exactly happened at these sites and areas outlined in this guidebook? What can we learn from these events? What are the problems still remaining? Give your five senses and imagination full play and contemplate these issues. At the same time, question anew our lifestyle, our lives, and the way of our society.

With this guidebook in hand, stroll in areas such as downtown Minamata, around the factory, and fishing hamlets where many patients came from. You may be by yourself or with your friends. Look at your own life in a mirror called "Minamata".

This guidebook is filled with the passion of people who have long been addressing and fighting Minamata disease in different sectors and with different capacities. Walk with these people in Minamata, understand their passion, and you will, we trust, think about facing up to Minamata disease. Through the lives of people living in Minamata, each one of us is required to come face-to-face with the past and present realities of this catastrophe.

It has been about 13 years since this guidebook was issued as Booklet ⑤ in English and the city of Minamata has been changing. We are now publishing Booklet ⑰ focusing on the Minamata of today.

We hope this guidebook will provide you with new consciousness and opportunities to meet many people.

How to use the map

■ Wide area map (p.10)

You can grasp the whole view of the areas along the Shiranui Sea which were contaminated by mercury discharged from Chisso Corporation.

Map 1 shows Minamata City, Map 2 shows the Ashikita seashore, Map 3 shows Goshoura, Shishi-jima Island, and Map 4 shows Izumi City in Kagoshima Prefecture.

On the ria shoreline of Ashikita between Yatsushiro City and Minamata City, which was designated as Natural Park by Kumamoto Prefecture, there are many fishing ports including Imuta, Sugisako, Tano-ura, Umi-no-ura, Ushi-no-mizu, Fuku-ura, Oya, Egoshi, Fukuhama, Odomari, (Map2), Marushima, Yudo and Modo (Map1). And fishermen's hamlets are formed around these fishing ports. Visit the fishing ports, following the side road which extends from the main street along the coast.

In Goshoura-jima Island in Amakusa, on the other side of the shore, there are hamlets; Oura Motoura, Eboshi, Goshoura Port, and Arakuchi. Climbing the Karasu Pass, visitors can enjoy panoramic views of the Shiranui Sea (Map 3). To the south below, visitors can see Nagashima Island and Izumi City in Kagoshima Prefecture, Komenotsu Port, Nago Port, and Fukunoe Port, which are thriving on the fishing industry. (Map 4) A tour of the remote islands is recommended, using a water taxi.

■ Map1 ～ 4

Map 1 shows Chisso Corporation (present JNC Co.Ltd) in Minamata City, and buildings and places related to the Minamata disease case around the Minamata factory. Minamata Station (E-4), areas around the Chisso Minamata factory (F-4), reclaimed land, the Water Amenity Seawall area (E-2), Tsukinoura (D-3), Yudo (C-3) and Modo (B-2) all have many Minamata disease sufferers.

Map 2 shows Ashikita Beach, Tsunagi, Ashikita, Meshima, and Tanoura. Map 3 shows Goshoura, the Shishijima areas, and Map 4 shows Izumi City in Kagoshima Prefecture. If time allows, a stroll around these areas is recommended.

When you decide the destinations you want to visit, confirm the institutes or the spots related to Minamata disease, on Maps 1~4. Then read the commentary on each of them by using the indication of page numbers or alphabet shown on the map.

You will be able to get a hint of what is to be realized and what is to be considered after visiting the institutes and spots.

As for the inscription of Chisso Corporation

Chisso Corporation's Minamata factory split off in 2011, and is now named JNC Co. Ltd. This guidebook transcribes the former name, as it gives a description of the past.

Wide area map

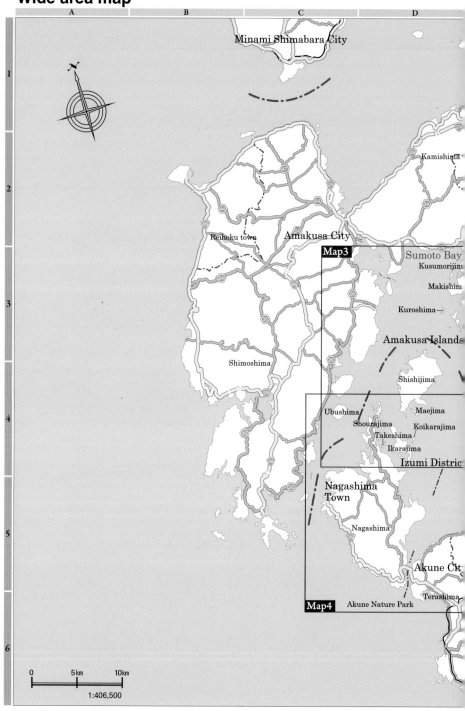

Minami Shimabara City

Kamishima

Reihoku town   Amakusa City

Map3   Sumoto Bay
Kusumorijim
Makishim

Kuroshima

Amakusa Islands

Shimoshima

Shishijima

Ubushima   Maejima
Shourajima Koikarajima
Takeshima
Ikarajima

Izumi Distric

Nagashima
Town

Nagashima

Akune Cit

Map4 Akune Nature Park   Terashima

0   5 km   10km

1:406,500

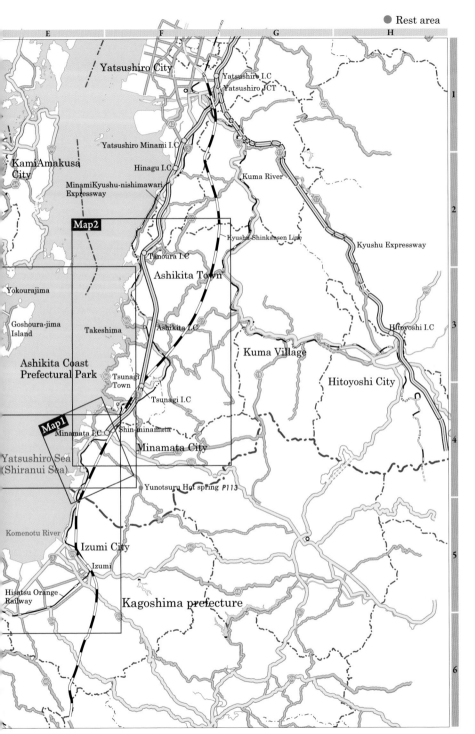

Map1　Minamata City

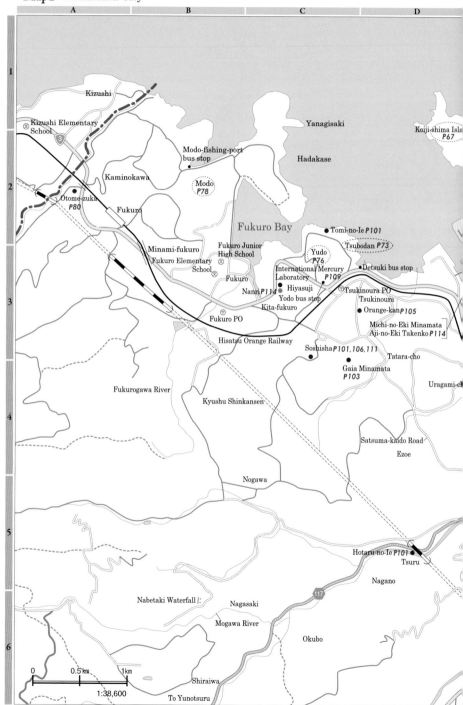

A B C D

1

Kizushi

⊗ Kizushi Elementary
School

Yanagisaki

Koiji-shima Isl.
P67

Modo-fishing-port
bus stop

Hadakase

2

Kaminokawa

Modo
P78

⊗ Otome-zuka
P80

Fukuro

Fukuro Bay

● Tomi-no-Ie *P101*

Tsubodan *P73*

Minami-fukuro
Fukuro Elementary
School ⊗

Fukuro Junior
High School

Yudo
P76

International Mercury
Laboratory
P109

●Detsuki bus stop

⑪ Tsukinoura PO

Tsukinoura

Fukuro

Nanri *P114* ● Hiyasuji

Yodo bus stop

● Orange-kan *P105*

3

Kita-fukuro

Ⓟ Fukuro PO

Michi-no-Eki Minamata
Aji-no-Eki Takenko *P114*

Hisatsu Orange Railway

Soshisha *P101,106,111*

Tatara-cho

Gaia Minamata
P103

Uragami-ch

Fukurogawa River

Kyushu Shinkansen

Satsuma-kaido Road

Ezoe

4

Nogawa

5

Hotaru-no-Ie *P101* ●

Tsuru

Nagano

Nabetaki Waterfall

Nagasaki

Mogawa River

Okubo

6

0   0.5 km   1km

Shiraiwa

1:38,600

To Yunotsuru

1

2

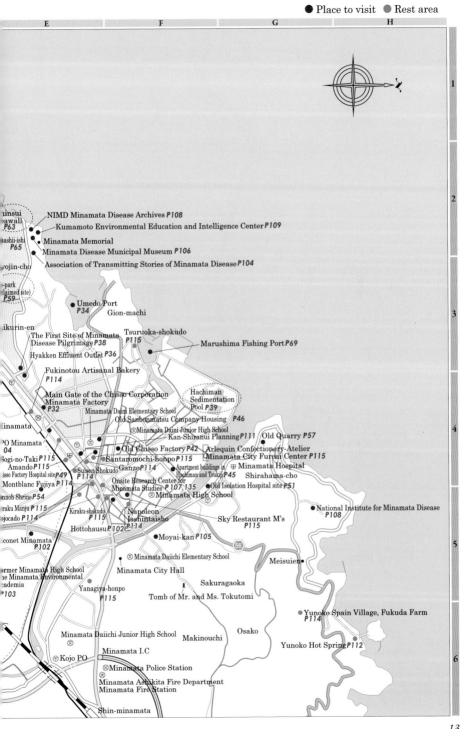

insui
awall
P63

NIMD Minamata Disease Archives *P108*
Kumamoto Environmental Education and Intelligence Center *P109*

aashii-ishi
P65
Minamata Memorial
Minamata Disease Municipal Museum *P106*
ojin-cho
Association of Transmitting Stories of Minamata Disease *P104*

-park
aimed site)
P59

Umedo Port
P34   Gion-machi

3

ikurin-en
The First Site of Minamata   Tsuruoka-shokudo
Disease Pilgrimage *P38*   *P115*
Hyakken Effluent Outlet *P36*   Marushima Fishing Port *P69*

Fukinotou Artisanal Bakery
P114

Main Gate of the Chisso Corporation   Hachiman
Minamata Factory   Sedimentation
P32   Minamata Daini Elementary School   Pool *P39*

inamata   Old Sanbonmatsu Company Housing *P46*
⊗Minamata Daini Junior High School
Kan-Shiranui Planning *P111* Old Quarry *P57*

4

O Minamata
04
ogi-no-Taki *P115*   Old Chisso Factory *P42* Arlequin Confectionery-Atelier
Amando *P115*   Santaromochi-honpo *P115*   ⊕ Minamata Hospital
isso Factory Hospital site *P49* Suisen Shokudo Ganzo *P114* Minamata City Fureai Center *P115*
P114   Apartment buildings in
Montblanc Fujiya *P114*   Hachiman and Tsukiji *P45* Shirahama-cho
nnoh Shrine *P54*   Onsite Research Center for   Old Isolation Hospital site *P51*
Minamata Studies *P107,135*
raku Manju *P115*   ⊗Minamata High School
jocado *P114*
Kiraku-shokudo   National Institute for Minamata Disease
P115   Napoleon   *P108*
conet Minamata   Isshintaisho
P102   Hottohausu *P102* *P114*   Sky Restaurant M's
P115

5

Moyai-kan *P105*

rmer Minamata High School   ⊗ Minamata Daiichi Elementary School
e Minamata Environmental   Meisuien●
cademia   Yanagiya-honpo   Minamata City Hall
103   *P115*
Sakuragaoka
Tomb of Mr. and Ms. Tokutomi

●Yunoko Spain Village, Fukuda Farm
P114

Minamata Daiichi Junior High School Makinouchi   Osako
⊗
Minamata I.C   Yunoko Hot Spring *P112*

6

⊗ Kojo PO   Minamata I.C

⊗Minamata Police Station
Minamata Ashikita Fire Department
Minamata Fire Station

Shin-minamata

13

Map2 Ashikita Coast

● Place to visit   ● Rest area

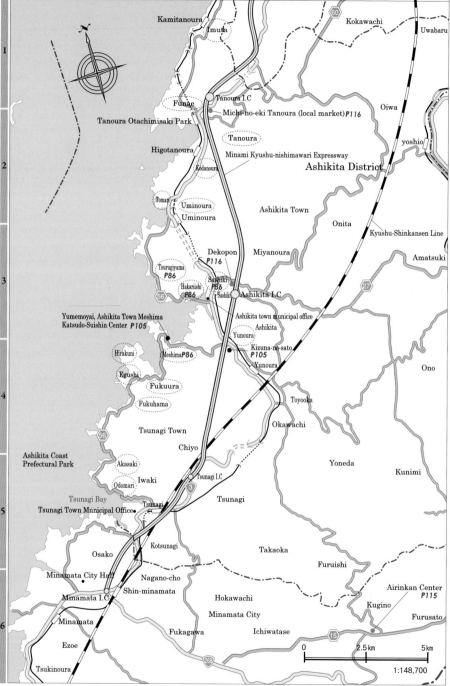

14

Map3 Goshoura, Shishijima

● Place to visit   ● Rest area

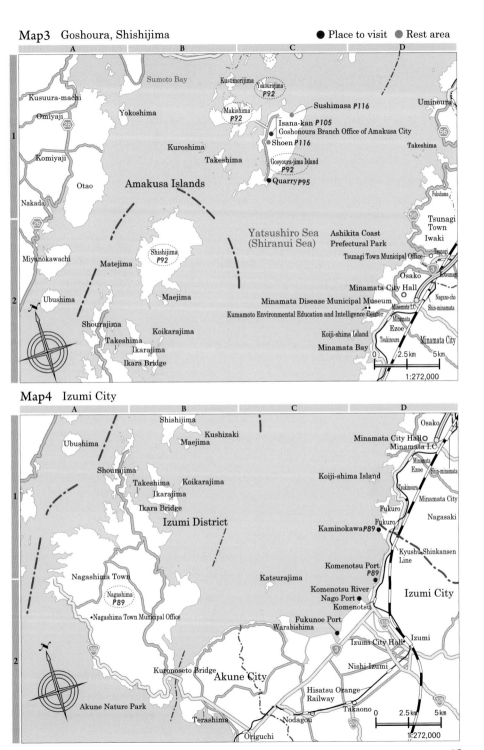

Map3 Goshoura, Shishijima

Sumoto Bay
Kusuura-machi
Omiyaji
Kusumorijima
Yokourajima *P92*
Makishima *P92*
Sushimasa *P116*
Uminoura
Yokoshima
Isana-kan *P105*
Goshonoura Branch Office of Amakusa City
Komiyaji
Kuroshima
Shoen *P116*
Takeshima
Takeshima
Gosyoura-jima Island *P92*
Otao
Amakusa Islands
Quarry *P95*
Fukuhama
Nakada
Yatsushiro Sea
(Shiranui Sea)
Ashikita Coast
Prefectural Park
Tsunagi Town
Iwaki
Tsunagi
Miyanokawachi
Matejima
Shishijima *P92*
Tsunagi Town Municipal Office
Osako
Ubushima
Maejima
Minamata City Hall
Nagano-cho
Shin-minamata
Minamata Disease Municipal Museum
Minamata I.C.
Kumamoto Environmental Education and Intelligence Center
Minamata
Shourajima
Koikarajima
Ezoe
Minamata City
Takeshima
Koiji-shima Island
Tsukinoura
Ikarajima
Minamata Bay
Ikara Bridge
0   2.5 km   5km
1:272,000

Map4 Izumi City

Shishijima
Osako
Kushizaki
Ubushima
Maejima
Minamata City Hall
Minamata I.C.
Shourajima
Minamata
Ezoe
Shin-minamata
Takeshima
Koikarajima
Koiji-shima Island
Tsukinoura
Ikarajima
Minamata City
Ikara Bridge
Izumi District
Fukuro
Fukuro
Nagasaki
Kaminokawa *P89*
Nagashima Town
Kyushu Shinkansen Line
Katsurajima
Komenotsu Port *P89*
Nagashima *P89*
Komenotsu River
Izumi City
Nago Port
Nagashima Town Municipal Office
Komenotsu
Fukunoe Port
Warabishima
Izumi City Hall
Izumi
Kuronoseto Bridge
Akune City
Nishi-Izumi
Akune Nature Park
Hisatsu Orange Railway
Takaono
0   2.5km   5km
Terashima
Nodagou
1:272,000
Origuchi

Outline of the Minamata Disease Case

<What is Minamata Disease>

Minamata disease was an environmental disruption on a large-scale that humans had never experienced before. A large amount of untreated industrial waste, which contained hazardous substances including organic mercury, was discharged into the Shiranui Sea. The effluent destroyed the environment, causing massive and serious damage to human health. Many Minamata disease patients have been suffering with various symptoms such as numbness in the limbs, visual field abnormality, ataxia, and so on. And the cases of Fetal Minamata disease were an unprecedented experience, in which the damage was caused through the placenta. The pollution and its destruction spread all over the Shiranui Sea, extending from Minamata Bay to Amakusa, to Nagashima in Kagoshima Prefecture to the islands on the opposite shore like Shishijima.

As of October 2020, there were 2,283 government certified Minamata disease patients (with an additional 715 in Niigata) and over 70,000 sufferers are receiving various medical benefits. In addition, the number of applicants for certification is over 1,000.

From the various points of view such as the number of damaged people, the affected area, and seriousness of the damage, it is an unprecedented pollution case.

<Outbreak and Expansion of Damage>

Pollution of the sea started before the war. The damage to the fishery had occurred frequently, and every time it occurred, Chisso paid compensation to fishermen as a solution. Local medical doctors took note of the outbreak of a disease with an unknown cause from the late 1940s. However, it was May 1, 1956 when the outbreak of Minamata disease was reported officially to the authorities.

The culprit, hazardous organic mercury, was discharged by the Chisso Corporation (currently JNC), who used it in their manufacturing process. At first, the cause was unknown, and the disease was suspected to be some type of epidemic. Soon, it was found to be noninfectious, and the company's effluent drew suspicion. By the autumn of the same year, a report was made, pointing out a heavy metal passed through fish and shellfish caused repercussions on the human body. At the end of 1959, Chisso concluded the notorious solatium agreement with patients through the mediation of the Governor of Kumamoto and the Minamata Disease Dispute Conciliation Committee, which stipulated that the patients should relinquish their claim to further compensation even if it was decided in the future that Minamata disease was caused by Chisso's release of effluents.

Chisso installed a pseudo purifier called a cyclater and spread propaganda touting the cleanness of its discharged water. Later it was exposed as having no effect at all. On the other hand, neither the national government nor Kumamoto prefecture took any countermeasures such as a fishing ban, a prohibition of eating fish and shellfish, or stopping the discharge of effluent from the beginning of the outbreak. Consequently, Chisso continued to discharge their effluent till 1968 and the damage expanded. In 1965, the second outbreak of Minamata disease appeared in the basin of the Agano River in Niigata, which resulted from the wastewater discharged by Showa Denko.

In 1968, the national government finally released its official opinion, confirming that Minamata disease was a pollution-related disease caused by effluent discharged from Chisso. Through this fact, those unknown disease patients were officially recognized as the sufferers of pollution. This was only the beginning of another long road of hardship for the sufferers of the disease. In 1969, patients filed lawsuit against Chisso. On March 20, 1973, the district court awarded a full victory to the patients, condemning Chisso's act of misconduct.

Then, with independent negotiations, a compensation agreement was concluded between the patients and Chisso. After this, government-certified patients were to receive a lump sum payment of 16,000,000 yen to 18,000,000 yen, a pension, coverage of medical costs, care costs, and so on.

<Expansion of sufferers and movement of patients>

However, due to the expanded pollution to the whole coastal area of the Shiranui Sea, the plaintiffs were not the only Minamata disease patients, but more and more patients filed their applications for certification. Nevertheless, national and prefectural governments set unreasonable and narrow criteria such as "Certification Criteria for Acquired Minamata Disease" (a notification by the Director General of the Environmental Health Department in 1977) for certification against the reality of the sufferings, which increased the number of uncertified patients, forcing them to negotiate directly and file further lawsuits.

In 1995, the patients' group and plaintiffs, except for the Kansai case plaintiffs, reached an out-of-court settlement, which began the government's relief measures. About 10,000 people received a lump sum payment (2,600,000 yen) and medical relief.

However, on October 15, 2004, the Kansai Supreme Court ruled in the Minamata Disease Kansai Lawsuit that both national and Kumamoto governments were responsible as same as Chisso for the first time in the history of Minamata disease. The plaintiffs in this case were those who did not respond to the reconciliation. At the same time, those plaintiffs who were the uncertified patients were acknowledged as Minamata disease patients as well to receive compensation.

After this Supreme Court decision, sufferers who could not raise their voices started to seek relief. The number of applicants for certification was over six thousand, and the number of Health Notebooks issued, which was the medical relief system resumed by the government, was over 20,000.

These applicants mostly had sensory disorder, which is peculiar symptom of Minamata disease. Among them are sufferers whose symptoms were critical or fetal Minamata disease patients. In 2007, a lawsuit seeking state compensation was filed by fetal Minamata disease patients and sufferers who belong to the Shiranui' Patients Group.

Such facts have made the national government start new relief measures which feature **a)** a lump sum payment of 2,100,000 yen and medical relief under the Act on Special Measures Concerning Relief for

Victims of Minamata Disease and a Solution to the Problem of Minamata Disease and **b)** starting the procedure of splitting Chisso company.

Application for this relief measure was closed at the end of July 2012. The number of applicants for this relief measure was about 65,000. This huge number served to confirm the vast scale of the damage which had been wrought.

These facts show there are still many people who cannot come forward as they are suffering with Minamata disease because of strong discrimination and prejudice.

＜New certification criteria and relief of sufferers＞

The Supreme Court overturned the decision of Kumamoto Prefecture, which rejected Chie Mizoguchi's case in Minamata and F's case in Osaka in a lawsuit regarding the rejection of certification in April 2013. And the court decided the obligation for Minamata disease certification and ruled that one who has only sensory disorder can be certified with comprehensive inspection of mercury intake. So the court rejected the claim by the national government, which claimed the combination of symptoms are required for certification. In March 2014, the Ministry of Environment issued a new notice that summarized the "comprehensive investigation" for cases having no combination of symptoms, further stating that objective evidence was required material (which shows that person consumed fish and shellfish 10 years prior, mercury concentration in hair or urine, or existence of a family member who is a certified patient).

At the end of August 2014, a judgement result regarding relief measures based on the Act of Special Measures was announced, and 38,257 people in three prefectures - Kumamoto, Niigata and Kagoshima (32,244 were eligible for lump sum money) - were to be eligible for the benefit. However, about 10,000 people were rejected, and the reason was not announced. Since it is not considered an administrative measure, people cannot register objections both in Kumamoto and Kagoshima Prefectures.

In 2020, lawsuits against Chisso, the national and Kumamoto governments by the Minamata Disease Sufferers Mutual Aid Association, the third litigation of the Niigata Minamata disease case, the second No-more Minamata litigation (Kumamoto and Niigata) were ongoing.

<Learn from Minamata and construct a future without pollution>

Though 64 years have passed since the official recognition of Minamata disease, the struggle between sufferers' appeals and officialdom concerning certification and damage compensation is still continuing. This is the indication of the failure of the relief and compensation systems and policies. We hope you can make a good use of the experience of Minamata for construction of a future without pollution.

Conclusion of the Mercury Convention and Minamata

The Minamata Convention on Mercury was concluded in Kumamoto City on October 10, 2013 and came into effect on August 16, 2017. As of the end of December 2018, 128 countries and the EU signed the convention, and 120 countries have ratified it. Following the international practice, it should have been named the Kumamoto Convention after the conference venue. However, it was called the Minamata Convention in accordance with the speech by then Prime Minister Yukio Hatoyama, who participated at the Minamata disease victims' memorial service. Due to insufficient compensation and the unsettled issue of Minamata disease, international NGOs and Minamata disease sufferers' groups questioned the naming. So, it is internationally called the Mercury Convention.

Resulting from the World Mercury Assessment implemented in 2002, UNEP (UN Environment Program) determined the following: Mercury use in advanced countries has been reduced, but mercury emission into the air is increasing. Artisanal small-scale gold mining in developing countries continues to use mercury which is released into the air and water. No matter how small the level of mercury exposure, once entering humans through the food chain, it causes developmental disorders, infertility, and heart disease. Accumulating in whales and fish, it has a high environmental risk. A reduction of emissions is required. Therefore, regulation of mercury use has been investigated internationally.

The contents of the Mercury Convention are: ① Prohibition of new development of mercury mines, ② Abolishing the use of mercury in the chlorine alkali process within the defined deadline, ③ Export and import are not permitted unless the use is permitted with the consent of parties, ④ Abolishing mercury-added products in 9 fields by the end of 2020, ⑤ Reduction of mercury use in artisanal small-scale gold mining, ⑥ Reduction of mercury emissions into the air, water, and soil, ⑦ Specify and assess pollution sites and reduce the risks, ⑧ Establishment of an international committee (secretariat) to administrate promotion and compliance of the regulations, ⑨ Parties to the convention will develop national laws, prepare national implementation plans, and tighten regulations.

The convention is comprised of 25 articles and 5 annexes. As for the operational procedure, COP (Conference of the Parties) were held twice to discuss and craft guidelines. COP3 was held in November 2019.

Regarding control of pollution sites mentioned in the convention, Eco Park (under which heavily contaminated mercury sludge dredged from Minamata Bay lies) as well as Hachiman sedimentation pool are applicable. Huge amounts of mercury have been contained. If it is to be maintained semi-permanently, further countermeasures will be necessary because the revetment was designed with a durable life of fifty years. The City of Minamata is now planning to develop the reclaimed land offshore with nominal terms of development of the estuary of the Minamata River as countermeasure for mercury remaining in the former Hachiman sedimental pool, which is the final disposal site of Chisso's industrial waste.

Chisso（presently JNC）and the City of Minamata

It has been 114 years since the establishment of Chisso Corporation. The City of Minamata has developed into a modern industrial city along with the Chisso Minamata factory. The development of Chisso had seemed to promise a bright future for Minamata. However, 64 years ago, in May 1956, all of a sudden it took a harsh turn for the worse. The official recognition of the outbreak of Minamata disease was the beginning of a new chapter of its history.

At the time of its official recognition, Minamata disease was called "Minamata strange disease" due to its unknown cause. The health damage brought by it and Fetal Minamata disease was far beyond our imagination. No one was able to foresee its immeasurable and long-lasting effects on the local people.

The Minamata disease incident presented a variety of problems to the local people. One of them is how to accept Minamata disease. There is no option for a patient's family to choose. They can only live with Minamata disease on a daily basis. However, other people were forced to choose whether to accept and face the reality of Minamata disease or avoid it as if it didn't exist. It can be said that these 64 years was the history of conflict over the acceptance of Minamata disease.

The citizens' movement of changing the name of the disease was a symbol of this complex situation. Appeals to change the name of Minamata disease because it was a cause of prejudice and discrimination against people from Minamata was launched several times: 1968, 1973 and more. This was a manifestation of a sense of crisis among those who wished to avoid Minamata disease and defend Chisso.

In 1956, the population of Minamata was about 50,000（10,000 of them were said to be workers at the Chisso Minamata factory and their families）. Presently, the population has decreased to 25,000 and the Chisso Minamata factory has been significantly reduced due to factory reorganization.

However, few people know that this is the core plant of Chisso which produces cutting-edge products such as liquid crystal.

Chisso was split off under the Special Measures Law enacted in 2009, and transferred its business to JNC, which was newly established in 2011. Chisso has become its holding company. If the company sells its shares, the name of Chisso will disappear and JCN will have no relation with Minamata disease.

The scenery of Minamata Bay was drastically changed with the mercury sludge treatment project. The sea of Minamata would never recover as a fertile source of bounty. The huge area of reclaimed land is only the witness.

During the history of over half a century, countless things have been lost in Minamata. Among them, the experience of the Minamata disease incident has been the biggest legacy, and how to utilize it in local regeneration is the key point for the future of Minamata.

The figure of Minamata in old days

Here is a map revised in 1911 with measurement in 1901, before Chisso factory was established in Minamata.

Where the Minamata River and the Yude River meets figuring of X, there is Nakanoshima Island and it is the center of the town. The downtown area of present Minamata was paddy field and there was a salt field near the sea. The shape of estuary of the Minamata was quite different from the current one. The photo on page 39 shows Hachiman Sedimentation Pool and that used to be the tidal flat in the sea. The old shape of Minamata bay much before reclamation can be observed. The factory must have been constructed on the riverbank of the Minamata, however, it was not confirmed on the map.

In 1956, 48 years later, the outbreak of Minamata disease was officially recognized. At that time the current Minamata had been formed.

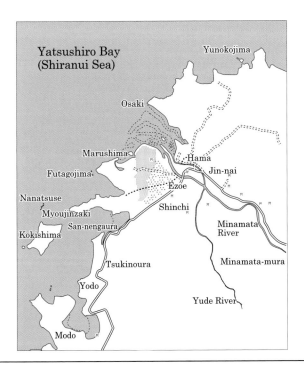

Compensation and Relief Systems for Minamata Disease

There are two systems of compensation and relief for Minamata disease compensation for certified Minamata disease patients and relief measures which provide certain medical assistance, but not recognizing sufferers as Minamata disease patients.

The former is the compensation system under the Law Concerning Special Measures for the Relief of Pollution-Related Health Damage (Law Concerning Pollution-Related Health Damage Compensation and other Measures since October 1973, here in after referred to as the "Ko-ken Law") enacted in December 1969. The Pollution-Related Health Damage Certification Council of Kumamoto Prefecture and Kagoshima Prefecture judge applicants and certify them as Minamata disease patients as determined by the prefectural governor. Once certified, a patient can receive either compensation based on the Compensation Agreement with Chisso in 1973 or compensation benefits provided in the Article 3 of the Ko-ken Law.

Compensation benefits under the Ko-ken Law are: 1. Medical treatment benefits and medical expenses, 2. Disability compensation, 3. Bereaved family compensation, 4. Bereaved family compensation lump sum, 5. Children compensation benefit, 6. Medical care benefit and 7. Funeral rites expenses. The amount is diverse in accordance with age, sex, degree of disability. The disability compensation in FY2018 was around 160,000 to 260,000 yen monthly. No one has yet received compensation benefits based on the Ko-ken Law in Minamata Disease.

The Compensation Agreement with Chisso includes consolation money, a lifetime special adjustment allowance, medical treatment benefits (hospitalization, commuting to hospital), medical expenses, nursing care expenses, funeral rites expenses, and expenses for acupuncture, moxibustion, medical allowance, hot spring recuperation, costs for diapers, costs for helpers, funeral offering, costs for massage treatment (up to 25 times a year), costs for schooling for fetal Minamata disease patients and traveling expenses to the hospital paid by Chisso. The consolation amount is determined in accordance with the ranking of symptoms, which is

16,000,000 to 18,000,000 yen, the special adjustment allowance is 71,000 to 177,000 yen monthly (as of 2015). Due to there being no indexation of consolation money, the amount has remained the same as it was when the agreement concluded. However, indexation will apply for others, so those amounts will change.

In 1996 the new political settlement was facilitated.

Those applicants who have not been certified and those whose reply from the certification committee or disposition of the governor is still pending, under certain necessary conditions, will be issued with a certification committee review medical notebook until the result is issued. The conditions are: residing in the appointed areas for more than five years and having waited at least one year since applying for certification (six months for those with severe symptoms). The benefits fully cover medical expenses and partially pay for nursing care costs – the patients must pay for the remainder. Moreover, up to five acupuncture and moxibustion treatments are covered per month although there is also a payment limit for these.

The latter is a relief measure under the Comprehensive Minamata Disease Medical Care Project (hereinafter referred to as Medical Project) and the Act on Special Measures Concerning Relief for Victims of Minamata disease and a Solution to the Problem of Minamata Disease (hereinafter referred to as the Special Relief Act).

In 2009, the Special Relief Act was established and transited to the "Minamata Disease Victim's Notebook". Under this Act there are two kinds of relief programs: one is to provide a lump sum payment of 2,100,000 yen and a medical service allowance resembling that of the Medical Care Notebook, and the other is to provide only a medical service allowance. Applications were received by July 2012.

Regarding the certification system for Minamata disease by the Ko-ken Law, the certification criteria were narrowed by "Certification Criteria for Acquired Minamata Disease" as a notification by the Director General of the Environmental Health Department (hereinafter referred to as the 1977 Criteria). Subsequently, more patients were rejected. On March

7, 2014, the notification by the Director General of the Environmental Policy Department, the "Consideration on Certification of Minamata Disease based on the Law Concerning Pollution-Related Health Damage" was issued. Due to "The contents of comprehensive consideration", the criteria had become more severe for the patients such as requiring organic mercury concentration in the body (concentration in hair, blood, urine and umbilical cord at the time).

The number of certified patients was 2,283 (including 1,918 dead) in Kumamoto and Kagoshima as of the end of December 2019. The number of Medical Care Notebook holders was 11,152, that of Health Notebook holders is 21,222, and that of New Health Notebook holders was 28,364 as of the end of May in 2010. That of Minamata Disease Victim's Notebook holders was 36,361 as of the end of August in 2014. More people are seeking relief, which means compensation and relief issues of Minamata disease have not been solved yet.

Status of certified patients in Kumamoto Prefecture and Kagoshima Prefecture (as of the end of December 2018)

	Certified (survivor)	Applicants
Kumamoto Pref.	1,789 (260)	727
Kagoshima Pref.	493 (88)	1,090
Total	2,282 (348)	1,817

Present Situation of Damage of Minamata Disease

When discussing Minamata disease, we tend to focus on individual health damage. That is because Minamata disease patients at the time of the outbreak were rather serious cases. And use at schools and in the media of images of patients severe convulsion was another cause. However, there were also social damages such as discrimination, prejudice, and rejection. We will see this point with the result of questionnaires which the Open Research Center for Minamata Studies carried out on 8,000 people with the collaboration of the Asahi Newspaper Company in 2016.

<Scale of Damages>

As for the numbers of sufferers of pollution or that of victims of natural disasters, quite accurate numbers or details of damage are officially announced. How is the case of Minamata disease?

From the official recognition of Minamata disease in May 1956 till the beginning of the 1970s, the number of sufferers was said to be 100 and some. Upon enforcement of relief measures such as Special Measures in 2009, more than 60,000 people applied. We can confirm that the total number of the damaged is about 70,000. This number is the total of individuals who made claims and became the subjects of reparation and compensation. If we include latent sufferers who were afraid of discrimination and uncertified deaths, this number will increase. Over 1,000 sufferers are applying for the certification now (as of the end of 2018).

<Regional spread of Damage>

The damage had spread all over the Shiranui Sea coastal area. Sea water moves around in the sea, and fish migrate. And it extended to mountainous areas where fish and shellfish polluted with organic mercury were carried along distribution routes. A dividing line can never be drawn to delineate the polluted area. However, this point also has been the disputed issue between sufferers and administration. Since the spread of pollution had never been surveyed except for the checking of mercury in hair of fishermen of coastal areas by Kumamoto and Kagoshima Prefectures, there is no strict method of checking. Only that estimation can be repeated.

<Damage and Suffering>

As for health damage, sensory disorders in extremity of the hands and feet, ataxia, difficulty maintaining balance, restriction of the visual field, gait disorder, articulation disorder, muscle weakness, tremor, disorder of ocular movement, hearing impairment, and others are the main symptoms. However, these are medical given names for the symptoms which the patients must bear, and they do not adequately communicate the true pain and difficulties which sufferers feel on a daily basis.

Subjective symptoms explain health damage. Many sufferers complain of numbness in the limbs and cramps. Numbness explains tingling, pain and lack of sensation. A cramp is not only pulling a calf muscle, but the pulling of a muscle of everywhere like limbs, back, and neck. It occurs at any time of day or night. There is no telling when it will strike or how long it will last.

The frequency order for other subjective symptoms is as follows; lack of sensation, lack of dexterity, trembling, fatigue, headache, difficulty in seeing around, difficulty in hearing, and frequent stumbles. Each of these are the difficulties they feel in their daily life.

Symptomatic treatment is the only way and there is no treatment for complete cure. Many patients and sufferers have been trying to reduce the pain with painkillers, acupuncture, moxibustion, and Chinese medicine.

<Social damage: discrimination and prejudice>

Distress is not only physical pain, but sufferers have mental anguish also. In the questionnaire, we asked what painful things they experienced after being stricken with Minamata disease. Of course, the most common answer was that they could not do any work or housework. A similar number of responders answered that there was anxiety regarding whether their siblings would contract Minamata disease or not. Those who answered "discrimination and prejudice" were 20 %. Then we asked about the Minamata disease-related harsh experience. About half of them answered that they were ridiculed, verbally assaulted, or become the subject of gossip. We had quite number of answers telling "they had difficulty in getting married, being refused dates, or were unable to find work."

We are astonished at the fact that Minamata disease sufferers could be the target of discrimination and prejudice. This is why about the half of sufferers only talk about Minamata disease with their immediate family (spouses and siblings), and only about 10 % talk with neighbors while 11 % had never talked about it with anyone else.

Reference: "the Final Report of 60 years after official recognition of Minamata disease questionnaire" issued by the Open Research Center for Minamata Studies in February 2019

Main Gate of the Chisso Corporation Minamata Factory
— Crossover Point of Prosperity and Resistance —

Map1 E-4

At 4 p.m. when the siren for shiftwork changes blared out in the factory, all the Chisso factory workers together began to go off work. The Minamata factory, with nearly 5,000 employees at one time, was located in the middle of downtown Minamata, and symbolized "prosperity". The Minamata Railway Station on Kagoshima Line, inaugurated in October 7, 1927, was built right across from Chisso's main gate. The factory and the Hisatsu Orange Railway Minamata Station still stand opposite from each other over Route 3.

Look at the factory from the main gate, and reflect on Chisso's turbulent history. It was through this gate that, in November 2, 1959, the fishermen of the Shiranui Sea forced their way into the factory to demand a suspension of the factory's wastewater discharge. On November 27, members of the Minamata patients' mutual aid society launched their sit-in campaign at the gate, which resulted in conclusion of the notorious

Sit-in by members of the Minamata disease mutual aid society (November 1959)

solatium contract (December 30). In 1962 during the dispute over a stable wage system, Chisso's union members formed a picket line at this same location. In 1971, patients calling for direct negotiations, led by Teruo Kawamoto and Takeharu Sato, started their 21 months long sit-in at this site. In 1988, patients again staged a sit-in, demanding "cause-effect adjudication". The most recent incident took place in October 2005, when Izumi group patients conducted a sit-in for apology and compensation. On every occasion, Chisso ignored the protesters. The entrance to the "Patients' Center" for the certified victims of Minamata disease is a small postern gate located around 50 m to the north of the main gate. Among the many entrances to the Chisso Minamata factory, the main gate symbolized the stony wall confronting the patients' protests. Chisso now accepts plant tours for school children and other visitors. When will Chisso open this main gate widely to the Minamata disease sufferers and citizens?

First day of the sit-in campaign by Teruo Kawamoto (October 1971)

Umedo Port (Used Exclusively by Chisso Corporation)
— Pivot of Marine and Land Transports —

Map1 E-3

Shitagau Noguchi, a founder of Chisso, was planning to build a carbide plant using electricity generated at Sogi power plant in Okuchi, Kagoshima Prefecture. His original choice of site was Komenotsu. Hearing this plan and consulting with some fellow villagers, Eitoku Maeda of Minamata invited Noguchi to set up the plant in his village by offering a low site price, donating electric polls necessary for power transmission, and advertising that Minamata had a natural good harbors in Umedo. Maeda's enticement was successful, and the factory was built in Minamata in 1908. Since the limestone used in the carbide plant was produced in Mt. Tsurugi near Sashiki and Himedo in Amakusa, Umedo was an ideal port for Chisso. As the factory was developed, so did the port. The liners from Amakusa and Misumi changed their anchorages from Marushima to Umedo. With the subsequent opening of Minamata Railway Station of Kagoshima Line, Umedo Port prospered as the pivot of marine and land traffic. In older days, coaches carried coals and salt to the gold mine in Okuchi, and on their return brought rice. Until just before the Amakusa liner shifted its anchorage to Hyakken Port, a horse-drawn omnibus service ran between Minamata Station and Marushima Port. Chisso's raw materials and products were carried between Umedo Port and the factory through a tunnel using initially trolleys. Later, another tunnel was built to use trucks and belt conveyors. In 1950's, ocean going freighters started to come to Umedo, taking the fertilizers to Okinawa and Taiwan. Roaring large cranes were operating to handle nonstop cargo loads of raw materials such as mineral phosphate and potassium chloride. In addition, the port had a thermal power station, and a billow of smoke constantly spewed out of its tall chimneys. Behind such prosperity of Chisso, fishing grounds of the sea were contaminated, and fishers as well as residents of Umedo fell ill with Minamata disease one after another. Later, the areas around Futago-jima

Island, rich in fish and shellfish, were reclaimed. At the time of the dispute over the stable wage system in 1962, Chisso's union members blocked the tunnels between the port and the factory. The fierce battles between labor and management frequently echoed across the port, with occasional police mobilization. At the time of the strife by coastal fishermen, those who were driven into a corner with Minamata disease staged a sea blockade of the port by roping boats together.

Umedo Port

Hyakken Effluent Outlet
― Ground Zero of Minamata Disease ―

Map1 E-4

From the onset of acetaldehyde production in 1932 until its termination in May 1968 (a result of the acknowledgement of pollution-triggered disease by the national government in September 26th of the same year), the Chisso Corporation continued to discharge its untreated wastewater containing methyl mercury into Minamata Bay. The maximum volume of mercury was believed to have amounted to a staggering 450 tons. This was indeed ground zero of Minamata disease.

In 1956 and 1957 when its effluent was increasingly suspected as the cause of Minamata disease, Chisso changed the location of its effluent outlet to the mouth of Minamata River. Between September 1958 and October 1959, the contamination spread throughout the entire Shiranui Sea area. A massive number of fish floated dead at the mouth of Minamata River; cats went mad and died in Goshoura; and Tsunagi Town adjacent to Minamata City saw outbreak of Minamata disease among its residents. Chisso committed what some have referred to as an "experiment on human bodies".

In 1957, Ministry of Health and Welfare found that Minamata disease was caused by the consumption of large quantities of fish and shellfish living in Minamata Bay. Yet the ministry did not apply the Food

Present Hyakken Effluent Outlet. The white building behind is a pump house. (Photographed in 2014)

Sanitation Act due to the lack of concrete evidence that all the fish and shellfish in the Bay were toxic. For this reason, the administrative action did not go so far as banning the fishing, but remained as a call for "voluntary restriction".

Before outbreak of Minamata disease, in 1926, fishers demanded compensation from the factory. People living by the sea felt through their day-to-day lives that the industrial effluent contaminated the sea and exerted an adverse impact on marine organisms. For example, they found that mooring their boat near the waste outlet could stop the fouling of the hulls of the boat because the barnacles did not attach themselves to the hull.

Minamata disease could have been prevented if the safety of wastewater had been confirmed before discharge. The damage of pollution would not have so widely expanded if the effluent had been suspended when it was found hazardous. There were several important moments, which include; 1957, when the Public Health Department of Kumamoto Prefecture announced that the cause of Minamata disease was within Chisso's effluent; 1958, when change of the outfall expanded the damage; and 1959, when experiments proved that cats developed the disease through exposure to the effluent. It is wholly reasonable that the accused here is not only the company which continued to drain the effluent without confirming its safety, but also the governments which failed to take any action even when the hazardousness of effluent was confirmed.

The First Site of Minamata Disease Pilgrimage   Map1 E-3

— From Shore of the Aga to Shiranui —

A jizo, or a stone statue as a guardian deity of children, is enshrined very close to the Hyakken outlet. It was a gift, given in 1994, from Niigata city to Minamata city. Mr. Teruo Kawamoto, who has been a mainstay in the campaign to help the Minamata disease victims, was hoping to build jizo statues at 88 locations in Minamata.

In Niigata, Showa Denko continued to discharge its untreated effluent with organic mercury into the Agano River. Consequently, many people in the Agano basin who lived on the wealth of river suffered from Minamata disease.

When people of Yasuda Town in Niigata heard of Mr. Kawamoto's wish, they went up to upper Agano to search for the right stone. The Agano River area produces excellent stones, and Yasuda Town has skilled masons. Carved and enshrined at Hyakken is the jizo statue made of stone from Agano River.

Four years later, the Yasuda people this time desired to erect a jizo statue beside Agano River made of Minamata stone. The stone for the Niigata statue was chosen from those of Minamata River that flows into Shiranui Sea. A stone with right size and shape was found and brought back to Niigata. With the hands of the same mason who carved the Hyakken statue, the stone from Minamata was turned into a jizo statue.

(From "Jizo Statue at Aga", published by Aganokai (Patients' Group of Aga) in September 2002)

The First Site of Minamata
Disease Pilgrimage

Hachiman Sedimentation Pool
— Carbide Residue and Dumping Site of Hazardous Substances —

Map1 F-4

With the Minamata River improvement project, a large portion of the shore called Chidorisu was reclaimed, and large-curved bank was built. Pre WWII, this area had salt fields and had been producing salt. Chisso, which was dumping the untreated carbide residue from the acetylene production into Hyakken Port, eyed the reclaimed land in Shiohama and the shallow beach of the cove near the mouth of the Minamata River. Around 1947, the company began to build a sea bottom reclamation pool using a stone embankment and concrete covering of the carbide residue landfill on a massive scale.

When the first sedimentation pool was filled, another one (Pool A) was developed farther off the coast of the shallow sea. When Pool A was filled, another one (Pool B) was developed near the estuary of the Minamata River. When Pool B was filled, another pool (North Hachiman Pool, 247,500 m^2) was developed where Chisso used to have their fields for salt production as well as their company housing. This was how Chisso expanded the landfills.

Around 1956, the company again faced the problem of disposing of carbide sludge, which was produced at a level of 200 tons a day. Their solution was a low-cost raising of the banks surrounding the vast but

Distant view of Hachiman
Sedimentation Pool
(Photographed in 2006)

39

already full pools which contained carbide residue.

With experiments on cats showing that the cause of Minamata disease was effluents from their factory, Chisso relocated the drainage channel from Hyakken outfall to these pools. Not only carbide sludge but also acetic wastewater containing mercury, sulfuric wastewater, and phosphoric wastewater were dumped into the pools. As a consequence, the causative substance of Minamata disease was flushed into and spread throughout the entire Shiranui Sea, which multiplied the number of Minamata disease sufferers. While the outbreak of mercury poisoning was contained to the areas around Minamata Bay, including Hyakken, Detsuki, Yudo, and Modo, September 1958 onwards saw the outbreak of patients in the Minamata River area, Funatsu, Hachiman, Yunoko, Tsunagi, and Meshima, reaching as far as Goshoura and Amakusa situated on the other side of the Shiranui Sea.

Subsequently, as a result of governmental administrative guidance and other pressures, Chisso began to pump the supernatant solution back to the sedimentation basin within the factory premise, and then the effluent was discharged from the Hyakken outfall until 1968 when the production of acetaldehyde was finally discontinued. The plants producing acetic acid and carbide have now discontinued their production. Part of the vast expanse of landfill containing toxic substances $(330,000 \text{ m}^2)$ was bought up by Minamata City to use as the waste incineration plant site. After the Great East Japan Earthquake on March 11th, 2011, the site has also been used as a solar power plant by making full use of a renewable energy purchasing system.

On the other hand, the first and the second landfills off Hachiman are used by Chisso as an intermediate treatment site and an inert-type sanitary landfill site. The entire ground here is nothing but an industrial waste disposal site.

    The revetment and roads which were under the management of Chisso were presented to Minamata City in 2005. However, the revetment has become old, so Minamata City plans to build port facilities around the area under the project called the Minamata River Mouth Promotion Project, and plans to reclaim the area next to Marushima New Port using earth from the construction of the Kyushu Motor Way.

    A huge volume of industrial waste containing mercury remains under the Hachiman Sedimentation Pools, and some people say that the sites should be kept under scrutiny as contamination sites based on the Minamata Convention.

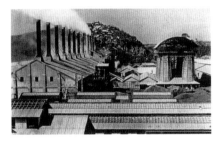

Old Minamata Carbide Factory

Canal of Hachiman Sediment Pool viewed from the entrance of Settsu Industries Co., Ltd (Photographed in 2006)

Old Chisso Factory
— Pioneer of the Modern Chemical Factory —

Map1 F-4

Before the Minamata River underwent improvement work, two rivers, one running through Ohzono and Koga and the other running through Jinnai and Makinouchi, used to cross in the shape of an X. Founded at a swampy paddy near the estuary in Koga side in 1908 was Nihon Carbide Shokai, or the Japan Carbide Company (later renamed Nippon Chisso Hiryo K.K, or the Japan Nitrogen Fertilizer Company). Joichi Fujiyama and Shitagau Noguchi, for the first time in Japan, started to manufacture lime nitrogen fertilizers and of carbide using the surplus power of the Sogi power station in Okuchi. This is the birthplace of what we now know as the Chisso Corporation (now called JNC). In 1909, a three-storied modern factory was constructed and was the third red brick building in Kumamoto Prefecture. People called it the old factory as opposed to the new one in current location. At present, only a part of it remains, but is a valuable heritage site.

With carbide, lime nitrogen, carbon, and iron plants, the old Chisso factory blazed a path for Japan's ammonium sulfate production, and was thus a pioneer Japanese modern chemical factory. Despite the smart-looking building with red brick walls, the company demanded much from its workers. The extremely hard labor and low wages drove off factory employees, which sometimes caused labor shortages. The old factory also had frequent explosions, causing a number of casualties and deaths. As a result, the company gained the reputation as a place where its workers would never reach old age. The company forced their Minamata laborers to work at an even lower wage than workers in other areas. Chisso's attitude was that 25 cents a day would do for Minamata workers 'because they have sweet potatoes to eat'.

Elders say that they worked 24 hours on and 24 hours off at first. Later it became two 12-hour shifts.

Workers at the carbon plant of the old Chisso factory were covered in carbon powder, with their entire body inky black except for their eyes, which were protected by primitive googles. When these laborers went to the public bathhouse, other workers would avoid them for fear they would be stained as black as the laborers. There is an interesting anecdote that, since the sparrows inhabiting in the carbon plant were all black, the workers of the old factory were sometimes called "carbon sparrows". The old factory established in Minamata did bring many benefits; it provided a workplace for people who no longer could work in agriculture and salt farms, and it introduced one of the earliest electric lights in Kyushu. At the same time, however, Chisso started to dominate Minamata.

Old Chisso Factory (Photographed in 2019)

On the other hand, Chisso' plan for expanding overseas started in the 1920s, and it built power plants in Korea and China. In 1927, it established the Korea Nitrogen Fertilizer Co. Ltd, and in 1930, Konan Factory started its operation. This was the foundation of the emerging zaibatsu, Nitchitsu Conglomerate.

Old Chisso factory was the third red-brick building in Kumamoto, and it is a precious industrial heritage, but it was not designated as a modernization period monument. The square factory building of red bricks was quite spacious, 50m each side, and many machines made by Siemens Germany were installed inside. In the factory premises, there were single-storied houses used as a main office and bathrooms. In the main office, a heating system with a traditional cooking stove was left. Keishi Isoda, a visiting professor of Kumamoto University who investigated the buildings, says that historical records concerning these buildings are lacking; therefore, further investigation is necessary. After Chisso let go of the buildings after the war, Egawa Co. Ltd has long used them for the purpose of their business. At the time of this writing, housing development is ongoing in the premises, but little remains of the red brick walls.

Chisso Company Housing (Detached House, Dormitory, and Apartment)
— Discrimination Based on Educational Qualification and Occupational Function —

Map1 E-4, F-4

As part of the welfare provisions, Chisso Corporation owned several company housing projects including detached houses, dormitories, and apartments within Minamata City. Length of service, age, family structure, and other factors were numerically rated to qualify for residency in the various housing projects. The point system for residency, however, differentiated, the layout of the house, the living environment, and the building structure, according to educational qualification and occupational function.

For example, the detached houses in Jinnai Housing with well-planned layouts and good living environments were allocated for those in senior management positions including the factory general director, department directors, and managers. On the property, there was a park and a sumptuous social club with recreation facilities, accommodation, and a banquet space for exchanges and business entertaining. They also hired household helpers. In the vicinity was a college-graduate-only dormitory for elites from prestigious universities in big cities, equipped with a pool and a hall for dances and playing table tennis. There was also a dormitory for singles, exclusively for high school graduates employed from outside of Minamata.

Deputy managers and group chiefs lived in an apartment or a detached house. A four-story apartment building was built in the

Hachiman housing before demolition (for group managers)

Map of Tsukiji Apartment area

Hachiman and Tsukiji areas at the lower Minamata River, whereas those who wished for a detached house were allowed to live in such housing.

Below them in rank were working foremen and unit heads, for whom the tenements with two or three housing units were allocated. The tenement houses had only a communal bathing facility; a bathroom was provided for each household in later years. Further below, the rank and file workers with fewer residency points were admitted to tenements with four or five housing units in Koga Housing and Sanbonmatsu Housing, again only equipped with a communal bathing facility. These tenement houses were more like huts, with thin walls which allowed little privacy.

Besides these housing projects, near the factory Chisso built corporate housing projects in places such as Hyakken, Marushima, and Shiohama so that their employees could be systematically and promptly mobilized in case of an accident or a disaster at the factory. After reexamining the old residency criteria, the company now promotes home ownership, and has demolished the old corporate housing. However, local Minamata laborers were rarely awarded residency.

After demolishing Hachiman Housing in 2010, Chisso established a mega-solar power plant with solar panels. (See p.47)

Jinnai Housing before demolition (for those in senior management positions)

Former Shiomi Housing (for deputy level managers)

Mega-Solar Power Plant and Minamata

Map1 F-4

When Chisso was established, its name was the Sogi Electric Co. Ltd, in 1906. Building a factory in Minamata in 1908, it started producing nitrogenous lime, using the method of the fixation of atmospheric nitrogen. During production, it built some hydro-power plants in Kumamoto Prefecture, including the Shirakawa Hydro-Power Plant in 1914.

Among the 13 hydro-power plants that Chisso built, 11 are in Kumamoto Prefecture, one is in Miyazaki Prefecture, and another one is in Kagoshima Prefecture, as of 2018, and all of them are controlled remotely by the Minamata factory. Its maximum output is 94,600kW, and the annual generation capacity is equivalent to satisfy 140,000 households, and a part of its power is sent to Chisso's Minamata factory through a 161km power grid.

Taking advantage of a renewable-energy-purchase system, which was established after the shutdown of nuclear power plants at the time of the East Japan Great Earthquake in March, 2011, Chisso built four solar power plants; in Minamata City, in Moriyama City in Shiga Prefecture, in Ichihara City in Chiba Prefecture. and in Kurashiki City in Okayama Prefecture, respectively. For its development, they used the knowledge they learned from hydro-generation plants, and these plants enable a maximum output of 16MW, with an annual capacity equivalent to satisfy 5,300 households.

In Minamata, demolishing Hachiman housing which was comprised of tenements of four or five partitioned units, Chisso built the Hachiman Solar Plant with a capacity of 2.6MW, and started its operation in 2014. Generally, a solar plant is supposed to be built on the site of reclaimed land or fallow fields where people don't live, but the JNC Mega-solar Plant is located in the center of a housing area, which creates a strange impression.

In addition, they built the Minamata City Shiohama Mega-Solar Power Plant at a final waste disposal site (at the site of former Hachiman sedimentation second pool), next to Marushima Shin-ko (new port), and started its operation in Nov. 2017, enabling a maximum output of 7.773MW.

Kyudenko Corporation and an Orix-affiliated company rent a site of sedimentation pool (as large as 87,000 m^2) from JNC under a contract of 20 years, and are selling power generated there to KEPCO.

Hachiman Solar Power Plant

Chisso Factory Hospital
― Director Hajime Hosokawa and his Cat Experiment ―

Map1 E-4

In downtown Minamata, the Suikosha Coop is located. (It was originally the coop for Chisso workers. After being the local coop for citizens, it now is the largest coop in Kumamoto Prefecture.) It is located, which used to boast the second highest sales per unit area after the Nada Coop in Kobe. Between the building and Minamata Daini Primary School on its west are a parking area, a bicycle parking space, a flower shop, a boutique, and other shops. The place of the coop used to be the site for Chisso factory hospital, which was closed in July 1969.

Originally established as a clinic for workers in Chisso, the factory hospital transformed itself in October 1948 from an in-house clinic to the first general hospital in Minamata, and provided medical care to many patients. Minamata at that time did not have a municipal hospital yet, and

Old Chisso factory hospital

the factory hospital recruited the best doctors and nurses from not only around Kyushu Island but also from all over Japan. Since the corporate hospital offered general practice together with an inpatients' ward, general practitioners in town referred patients beyond their capacity to the hospital, and consulted the hospital when they had questions. In this way, the annual total of outpatients reached 100,000 in some years.

Incidentally, in 1955, Chisso's corporate health insurance society had approximately 20,000 members, including subcontract workers, while the population of Minamata was 46,233.

The employees and subcontract workers of Chisso and their families relied mostly on the factory hospital for medical treatment. In case of explosions, accidents, or any emergency at the factory, the hospital's staff rushed to the scene rapidly to provide medical attention and necessary measures.

On April 21, 1956, a girl 5 years and 11 months old was brought to the pediatric department of the factory hospital, and her chief symptoms were brain symptoms. Her younger sister was also hospitalized with similar symptoms on April 29. Their mother complained that there were more patients in her neighborhood suffering from the same symptoms as her children. Hearing this information, the director of the factory hospital at the time, Dr. Hajime Hosokawa, reported to the Minamata Public Health Office (directed by Hasuo Itoh) on May 1, "There was an epidemic of an unknown disease of the central nervous system." In later years, this date was recognized as the official discovery day of Minamata disease. Chisso factory hospital was known for the humane practice of Dr. Hosokawa. However, its name is more closely associated for the so-called Minamata disease cat number 400 experiment.

The Isolation Hospital
― Starting Point of Minamata Disease Discrimination ―

Map1 F-4

Driving on the coastal road from Minamata Ohashi Bridge near the mouth of the Minamata River to the Yunoko area, you will see the Minamata City Senior Citizens' Welfare Center (completed in 1974), Keiai-en, the site of an old people's home, a school lunch center, and Minamata City municipal facilities. This is where the former Minamata City Hospital for Isolation was established. It was built after the Seinan Civil War in 1877, when cholera spread around the area. The Minamata Village Hospital for Isolation was opened here in June 1890. As the hospital was located at the river mouth of the Minamata River, the layout of the area changed largely after the reclamation projects were provided, but the hospital was located at the extreme edge of the town. In the postwar period, it was mainly patients with dysentery who were admitted here. And this is where those with Minamata disease were also admitted.

In 1956, two girls of five years eleven months old and two years eleven months old developed Minamata disease, and later many other patients appeared. After conferring with public health office authorities and factory hospital doctors, the Minamata Strange Disease Committee

Image of then isolation hospital (by Kenjiroh Kojima)

decided to admit them to the isolation hospital, in order to eliminate the anxiety of citizens and other hospitalized non-Minamata disease patients. Consequently, they started to repair the damaged isolation hospital. After the repairs were completed on the 27th of July, 18 patients of Minamata disease who had been admitted to the Chisso Factory Hospital were taken to the isolation hospital secretly, after diagnosing their disease as "pseudo Japanese encephalitis", in accordance with the Infectious Disease Prevention Law. Dr. Hosokawa and Dr. Kaneki Noda, pediatricians at the Chisso Factory Hospital had diagnosed the patients and judged by their cerebrospinal fluid test and infection route that the disease was non-infectious. They thought it might have been contracted from fish in the ocean, but there was no conclusive evidence to support this, and accordingly they could do little but diagnose it as an "unknown disease".

The hospital premises were covered with big thick pine trees, which gave people the impression that it was a scary place. Few people inhabited the area beyond the hospital. The atmosphere of the Shirahama area where the hospital was located is illustrated in the novels "Kugai jyodo" by Michiko Ishimure. The illustrated sketch in this booklet is depicted by Mr. Tsutomu Matsumoto, who was involved in the repair of the hospital when he was working as a staff member of the Construction Division of Minamata City. There were 3 buildings in the premises. One was for the patients displaying severe symptoms, another one was for patients with light symptoms. Patients were classified according to the severity of their symptoms. In the building deep inside, there was a room for corpses and an incinerator to burn the patients' clothes. Two patients were admitted in one room, and each room had windows on both sides for ventilation. Patients' families were responsible for their care as well as the provision of meals. Doctors were not always stationed there, and Dr.

Hosokawa from Chisso factory hospital is said to have come by regularly to examine patients.

"Two children were transferred to the contagion ward in Shirahama. Following this, neighbors abruptly changed their attitude toward the family. People who used to visit them suddenly stopped coming." (According to Asao Tanaka) Houses of patients were thoroughly disinfected by local municipal staff belonging to the Sanitary Affairs Division, based on the Infections Disease Prevention Law. Minamata disease patients were isolated in the isolation hospital until the end of August of the same year. Afterwards, they were transferred to the Fujisaki-dai Branch Hospital of Kumamoto University. The action of disinfection gave people the groundless impression that Minamata disease was infectious. It stimulated the anxiety of local residents, and undoubtedly it spurred discrimination again Minamata disease sufferers.

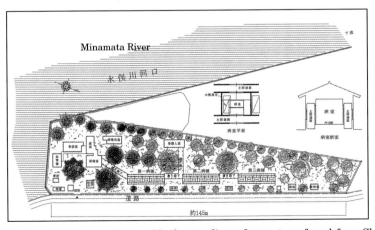

Isolation hospital where patients of "unknown disease" were transferred from Chisso factory hospital (Sketch S : 1/750)
Source: Collection from Open Research Center for Minamata Studies, Kumamoto Gakuen University

Vista from above Sannoh Shrine
— Enjoying the 180-degree Panoramic View of the Southside of Minamata City —

Map 1 E-5

Heading toward Yunotsuru Spa from the city center, you can see a bridge called Enan-bashi on your left after Hira-machi. Keep straight on along the road for 100 m (without crossing the bridge), and a fork in the road will appear. Take the road to the right and climb up for 700 m, and you will find Sanno (official name: Ezoe) Shrine.

There is a stairway between the shrine and the road. At around 30 steps up, a guardian deity called Agumasan is enshrined to protect the entire Hira-machi area from fire.

This spot, from the side of the small stone chapel, commands a 180-degree view of downtown Minamata.

From West-southwest to West-northwest

Clearly seen in the left of the picture is Koiji-sima Island. In front of the island is the area of Minamata Bay reclaimed land and the Hyakken drainage outlet. Nagashima Island in Kagoshima Prefecture can be seen across the sea.
The station in the middle of the photograph is Minamata Station of the Hisatsu Orange Railway, and nearby lies the main gate of Chisso Corporation and its factory.
Ikara-jima Island (in Kagoshima Prefecture) can be observed over the sea.

The far left, to the west-southwest, shows Koiji-shima Island, the reclaimed land at Minamata Bay, and other places. Turning gradually right, you can see the Chisso factory (present name JNC) and Minamata Station of the Hisatsu Orange Railway. To the right, although only through the gaps between trees behind the shrine, a view of the heart of downtown Minamata and the area around Minamata estuary can be enjoyed. Further toward the right, you can see on the right Chisso's Jinnai housing, and the building of Minamata High School (It was closed in March 2014 due to school consolidation). A part of the school building is used by "Environmental Academia." It is a place well worth paying a visit if time allows.

From Northwest to North-northwest

The left-hand side of the photograph is northwest (toward Marushima fishing port). The center of downtown Minamata can be seen through the tall trees behind the shrine. The island across the sea is Shishijima (Kagoshima Prefecture). The center of the picture is north-northwest (toward the Minamata River mouth), with places such as the site of the Hachiman company housing, the site of the old Hachiman Sedimentation Pools, the Old Quarry, the site of the Old Isolation Hospital, and the Onsite Research Center for Minamata Studies.
Goshoura-jima Island (Amakusa) can be seen far across the sea.

From North-northeast to East-northeast

In the north-northwest, the Jinnai district can be observed. At the point where the Minamata and Yude Rivers meet lies Chisso's Kosaki pump station. The name Agumasan came from the mountain opposite from it, Akiba-san (Mt. Akiba). In the east-northeast, Chisso's Jinnai housing can be observed. The building on the far right corner is "Environmental Academia", (formerly Minamata High School).
In the distance, the mountains in Tsunagi and Ashikita Towns can be seen one over the other.

Vista from the Old Quarry
— Enjoying the Extensive View of Downtown Minamata from the North —

Map1 G-4

Driving on the coastal road toward Yunoko Spa, one of the hot springs in Minamata, you will see Yasuragien (a healthcare facility for the elderly) and Minamata Hospital. Just ahead on the right along the road is the site of what used to be a quarry. Pass over the quarry site, and you will find Ohsakigahana Park on your left, and then the golden inscription for the "Monument of Gratitude". About 200 m beyond the monument, there is a fork in the road. Take the right-hand uphill path and proceed for about 600 m, and you will reach the upper part of the quarry. This point commands a view of Minamata Clean Center (on the site of the old Hachiman Sedimentation Pools, see p.39) and the mouth of the Minamata River.

In 1958, Chisso began to discharge its wastewater with acetaldehyde into the estuary of the Minamata River via the Hachiman sedimentation pool. The shifted location of the outfall combined with the effects of sea currents is considered to have multiplied the damage, with new patients appearing in the Hachiman, Shirahama, Yunoko, and Tsunagi areas.

Southeast

Chisso's Tsukiji apartment is seen in the front, and Hachiman housing in the back. (Photographed in 2006)

In the southeast, you can see Chisso's Hachiman housing and four-storied corporate apartment in Tsukiji, which are both described on page 45.

Although highly recommended, this spot becomes hard to access in summer because the path to the upper quarry bristles with weeds. What with the dense vegetation and a rather confusing entrance of the path, it is advisable to visit there with a local guide.

South

The building with chimneys just across the river is the Minamata City Clean Center. It is an incineration plant for household garbage. Several white buildings behind its right-hand side are factories and offices in what is called Eco Town. All these buildings stand on the site of the old Hachiman Sedimentation Pool, the landfill for Chisso's industrial wastes (carbide sludge landfill site).

A smoking chimney can be seen far beyond the Clean Center. It belongs to the Chisso Minamata Factory.

Behind a small hill to the right rear of the factory lies Umedo Port, which is exclusively used by Chisso. Beyond it is the reclaimed land at Minamata Bay.

West

Looking toward the west, you will find Chisso's (present name JNC) industrial waste-related facilities, such as the final disposal site (inert type sanitary landfill site) and the incineration plant (the building at the tip of the river mouth).

The wharf just above the center of the picture is Marushima fishing port.

Visible far across the water are the islands of Nagashima and Shishijima, which both lie in Kagoshima Prefecture (Photographed in 2006).

Eco-Park Minamata (Minamata Bay Reclaimed Site)
— The Buried Mercury Sludge —

Map1 E-2, E-3

In 1968, the national government recognized that the cause of Minamta disease was the effluent from Chisso. In 1970, the Kumamoto Prefectural Government asked Kumamoto University to consider methods for sludge treatment, but the university came to the conclusion that there was no effective solution. After a judgement was made in favor of the patients in the first lawsuit in 1973, the Kumamoto Prefectural government, Ministry of Environment, and Ministry of Transport discussed the matter, and as a result, the port management body (Kumamoto Prefectural government) entrusted the national government (Ministry of Transport) to carry out the Pollution Prevention Project and Port Development Project, which were started in 1977. In December, however, the court approved the temporary injunction filed by approximately 2,000 coastal residents including Teruo Kawamoto and other Minamata disease patients to suspend the sludge dredging work. The operation was interrupted until its resumption in June 1980.

Cylindrical and semi-cylindrical cells composed of cement and steel sheet piles were driven into the sea bottom. Sand was then injected into these cells to construct a bank revetment. In 1973, the Kumamoto Prefectural government divided the sea area of Minamata Bay into 500 m meshes, and surveyed the total mercury concentration of the bottom sludge. Based on the result of the survey, the sludge with more than 25 ppm was vacuumed up by suction dredger from the area through a suction mouth (resembling a large vacuum cleaner) and was taken and discharged into the landfill site.

The catch of contaminated fish, carried out by fishery compensation, was packed in oil drums, and together with sludge they were covered with synthetic fiber fabric (sheets), a grid of ropes, and watered white sand, in that order. Lastly, a one-meter-thick mountain of soil coated the layer, which

59

completed the soil covering work. Most of the finishing soil came from the quarry on Goshoura-jima Island located to the northwest of Minamata City. The project was completed in March 1990, costing 48.5 billion yen in total.

The reclaimed area is developed as a sport park, called Eco-Park Minamata, and utilized by citizens.

Underneath the 58.2 ha landfill site, approximately 1,510,000 m^3 of sludge containing over 25 ppm mercury is buried. A bamboo garden was laid down right next to the Hyakken effluent outlet, the ground zero of Minamata disease. From there all the way to the Shinsui seawall, Eco-Park Minamata spreads out to conceal what is a "massive industrial waste dump site".

According to Article 12 of the International Conventions on Mercury, which concluded in October 2013 and took effect in August 2017, the sites contaminated by mercury have to be identified and then their risk to the environment have to be assessed, and repaired, if necessary. Bank revetments composed of cement and steel sheet piles should be durable for 50 years. And in order to maintain it permanently, repair work is necessary every decade. There is a risk that steel sheet piles might be corroded by seawater, bank revetments might collapse in an earthquake, and mercury underneath might leak through liquefaction.

The Kumamoto Prefectural Government had a meeting convened in 2008 to discuss issues of earthquake resistance and deterioration of the reclaimed site constructed under the Minamata Bay Pollution Prevention Project, and compiled a resultant report in February 2015. According to the report, the bank revetment is thought to be safe until 2050, and specific measures will be discussed after convening the committee in 20 years. The Kumamoto Prefectural Government established a committee to maintain the reclaimed site and bank revetment under the Minamata Bay Pollution

Prevention Project in March 2016, and the site is controlled by this committee, but the survey was not provided even after the Kumamoto Earthquake.

The current mercury lying underneath Eco-Park Minamta must be supervised permanently. In order to mitigate the burden on the people in the next generation, some say that reclaimed mercury has to be collected, decomposed, and the sludge underneath has to be purified by employing technique widely practiced in soil contamination countermeasures.

Eco-Park Minamata and Koiji-shima Island（Photographed in 2013）

Containment Net

— Containing the Polluted Fish and Publicly Advertising the Safety of Fish Outside the Net —

Map1 D-2

In January 1974, Kumamoto Prefecture set up the containment nets to stop the polluted fish in Minamata Bay from scattering. The total length was 2,350 m. With commencement of the pollution prevention project in October 1977, the containment area was extended toward the area surrounding Koiji-shima Island. Nevertheless, the mesh of the net was 4.5 cm by 4.5 cm, large enough for small fish to swim through. As a passage for boats, there is an open area without the net (223 m wide). Acoustic devices were installed to prevent coming and going of fish, although a fish sitting right on the device was photographed. While described at the time of installation that this system would completely isolate the contaminated fish, the state and the prefectural governments later admitted that the effectiveness of the net was somewhere between 60 and 70 %. In 1990, after the reclamation project was over, the prefectural and city authorities as well as the local fishery cooperative put up signs to deter angling within the bay. On the weekend, however, an array of fishing rods was observed at the quay of the bay. In June 1995, the 2,298 m nets in Nanatsuse waters were removed after the

government claimed that intensive captures cleared out all the toxic fish. Furthermore, when the seven designated fish, such as rockfish, showed the mercury levels lower than the national limits, the Kumamoto Prefecture declared the fish and shellfish of Minamata Bay to be "safe". Following the declaration, they set out to remove the 2,106 m nets at the heart of bay, which was completed in October of the same year. With this hasty ending, the containment net which had symbolized Minamata disease constantly for over 20 years was gone, and the damages caused by Minamata disease were yet again obscured.

Containment net
allowing fish to pass
through

Shinsui Seawall (Shinsui Green Area)
― Edge of the Reclaimed Land where Organic Mercury Rests ―

Map1 E-2

The Shinsui seawall, completed in March 1994, is located on the very western edge of the Minamata Bay reclamation land (see p.61).

The seawall made of natural stone is shaped in a staircase pattern so that visitors can touch the seawater at high tide. Right beside the seawall is a 465 m esplanade with an eco-friendly boardwalk made without artificial preservatives. Along the promenade stands a stone statue (created by Hongan-no-kai or "group for salvation") praying toward the sea, and the Forest raised from seeds with planted young trees.

The pamphlets published by the city and the prefecture described this seawall area as "a place where people can have a walk and enjoy the blue of the sea, the green of Koiji-shima Island, and the view of the islands of Shiranui Sea" and "an amenity space where tourists and residents alike

Shinsui Seawall

can stroll while enjoying the sound of waves and the scent of sea breeze". Despite these attractive phrases, we should never forget the facts. This is the rim of the landfill site developed because of fears that the sealed dredge could spill out and pollute once again. In this sense, this is the boundary between the natural sea and the artificial land where organic mercury, the causative substance of Minamata disease, is buried.

As a place for prayer, some events associated with such purposes have been held here. Since 1994, The "Fire Festival" (sponsored by Minamata City) has been organized to offer prayers to all the victims of Minamata disease and all the living things buried in the reclaimed land. The Minamata Disease Memorial Monument was erected in 2006 to commemorate 50 years of official recognition of the Minamata disease outbreak. On May 1st every year, a memorial service is held. Visitors are encouraged to view the formerly fertile ocean, recalling the past experiences of Minamata, standing where mercury sludge remains below as part of the reclaimed land.

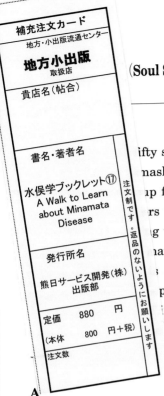

(Soul Stone) by Hongan-no-kai (group for salvation)
— Statues that Continuously Appeal —

Map1 E-2

ifty stone statues facing seaward in the Shinsui green
nashii-ishi, or "soul stones" made by the members of
ip for salvation", headed by the late Mr. Yoshiharu
rs placed their thoughts and their prayers into the
g to pass on the memories of pain and sadness of
many people.

; 30 cm square by 60 cm tall, carved with each
pes, and prayers. Showing various expressions, a
t were constantly appealing never to forget the
l potentials, and the loss caused by Minamata

ai Goup

· "group for salvation", was established in March
19 ___ initiative of 17 patients including Mr. Yoshiharu Tanoue, Mr.
Tsuginori Hamamoto, Ms. Eiko Sugimoto, and Mr. Masato Ogata. Their
anguish and yearning remains in Hongan-no-sho ("book of salvation").

After the long litigation battles and
compensation discussions, time drifted by and we
were just about facing the end by accepting the
settlement. At this time, Mr. Tanoue made an
appeal from the bottom of his heart: "The current
situation is forcing patients to die like dogs. All the
costs we paid and all the sacrifices we are
continuously compelled to make, they will be
entirely forgotten when the patients are all gone.
How on earth can we possibly leave proof of our life
in this world?" This statement led to the creation of

Stone statue erected by
Takeshi Sugimoto

our group. Later, we focused our efforts on foresting the landfill area of Minamata Bay, erecting jizo statues (soul stones in the green space, and reviewing closely the modern world we live in. Our activities still go on. (From Tamashii Utsure [Move in, Soul] , published by Hongan-no-kai in November 1998)

Hongan-no-kai
Office: Junpei Kanazashi 0966-63-2980

Stone statue erected by Dr. Masazumi Harada

Koiji-shima Island
― Legendary Island Affected by Minamata Disease ―

Map1 D-2

The island lying right ahead of the Shinsui seawall is Koiji-shima, also known as Koki-jima Island. On an old map, the island was marked as Koki-jima.

Koiji-shima Island (about 4 km around and 263,700 m^2 wide) used to have extensive pine forests. Though inhabited at some point, the island is now deserted. Soon after the war, in 1947, a shelter for 18 children called Nisui Kaiyoh Gakuen (predecessor of Hikari Douen Children's Home) was established to provide child-care works until its relocation two years later. Around 1950, Tsuya, mother of Minamata disease patient Yoshiharu Tanoue, lived here and was a kind of forest-ranger. In 1951, a city-run campsite was opened, bustling with beachgoers. As the impact of

Koiji-shima Island

Minamata disease expanded wider, the number of campers dwindled away year by year. The campsite was closed in 1959, and the island has been left uninhabited.

The capes of the islands have been providing valuable landmarks to the fishers, who called these points as "Harinomemenzu (Point Harinome-zaki), Point Kasase-zaki, and Point Kitambana. A lighthouse was also built in 1957.

The island was known for Tsumagoi-iwa (the stone of a wife longing for husband) on *Hiren Densetsu* [legend of tragic love] published in the Tensho era (between 1573 and 1591), so-called Koinoura Beach (lovers' beach), and Ebisu Shrine to pray for good hauls. In addition, a colony of *Machilus tunbergii* (a tree related to laurel, cinnamon and camphor trees) made the island famous. One of the last stands of the lauraceous (or 'aromatic bark and foilage') trees in Japan as a natural forest is found here, ironically due to the fact that the island was not subject to development due to the outbreak of Minamata disease. Needless to say, how we cherish Koiji-shima Island and its surround sea plays a key role in determining the course of Minamata's rehabilitation.

Marushima Fishing Port and Fish Market
— The Fish Could No Longer Be Caught —

Map1 F-3

The vicinity of Marushima Fishing Port located at the mouth of the Minamata River was a vast sandy beach, where seine fishing used to be practiced. The nearby Gamenkubi bathing beach was shallow to a considerable distance from the shore, being fertile with shellfish, including manila clams, common orient clams, tairagi fan-mussels, and mate shells. At low tide, the beach used to be busy with shellfish catchers, some of who came all the way from the mountain areas. Up to around 1950, sea bathers visited the beach in summer, some from Minamata and others even from Yatsushiro and Kumamoto.

The coast along Minamata is what is known as ria (deeply indented shoreline), and is abundant in a variety of fish. Traditional fishing hamlets in Minamata were the Funatsu (Hachiman) and Hama districts. After 1897, the center of fisheries in Minamata moved to Marushima. As the number of boats entering the port increased, an improvement plan for Marushima Fishing Port was promoted in 1937. Later on, Kumamoto Prefecture designated Marushima as a port under its management, and continually upgraded it as necessary.

From around 1953, coinciding with the Minamata disease outbreak, people saw stripped mullets floating and swimming dizzily near Marushima fishing port. When some fishermen developed Minamata disease, the fishery cooperative pressured fishermen and fish brokers never to apply for certification as Minamata disease patients, for fear that their fish would not sell because of the disease.

Meanwhile, fishmongers initially managed to sell the fish from Minamata by falsely labeling them. When it became increasingly difficult to sell fish, the fishmonger guild staged a boycott of fish caught in the seas off Minamata. Beleaguered fishermen were driven into debt, and they urged Chisso to compensate their livelihood, which developed into what we now know as the fishermens' dispute. In 1977, in addition to a reclamation project for Minamata Bay, the Kumamoto Prefectural government promoted the Pollution Prevention Project of Marushima Fishing Port, and a new port was established.

Marushima fishing port used to have two markets, one which was run by the fishery cooperative and the other by a private company. There is only one market now after the two merged into Shin Minamata Uoichiba K.K., or "New Minamata Fish Market Co., Ltd.". As typified by such, fisheries in Minamata suffered a heavy blow from the outbreak of Minamata disease. Now fishes from Izumi, Nagashima, and Shishijima in Kagoshima Prefecture, and Goshoura, Tsunagi, Ashikita, and Hinagu in Kumamoto Prefecture are unloaded here, but the port is not as busy as before.

Marushima Fishing Port at present (Photographed in 2014)

Column: Fish unloaded in Minamata

<Is fish in Minamata safe to eat? >

The question is difficult to answer in fact. Because it is as if the existence of fish shops is questioned. We cannot answer "yes", simply to this question. But what can be said now is that the risk is not at a substantial level.

I always answer so to this kind of question. Citizens in Minamata have got along with fish that was allegedly unsafe for 60 years. For some time, we didn't even know it was unsafe, but after knowing it was unsafe, we still get along with it in our own way.

For us, fish is as essential as air. Please understand the life of people living in seaside areas. All of us share the same fate. But as long as we are careful about the amount and frequency we eat fish, we don't have to worry so much. There is a saying that we should be frightened in a forward manner. What is important for us is to advance forward while knowing its risk.

< We realize that the taste of fish caught in Minamata is really good. Why is it so good? >

Whenever we visit other local sea areas, people there proudly say that fish caught in their sea area is the best. It is quite natural for them to say so, but I feel that the fish in Minamata (in the Shiranui Sea) is really exceptional. Compared to fish caught in other sea areas, the fish harvested here is clearly different. It is plump and has plenty of fat. My opinion is well-grounded. The secret of the taste comes from the type of water around Minamata.

The sea of Minamata is said to breed fish. Affluent spring water is here, and capes and islands are covered with evergreen forests (providing places to breed). It used to have wide ranging tidal flats (which are typical for deeply indented coasts, but most of which have now disappeared.) All together, they protect and breed many living creatures in our sea area. Nutritious water breeds good fish with a rich taste.

Furthermore, sea area of Minamata is highly closed, even in the closed inland sea area of the Shiranui Sea. Here, fishes at the top of the fish chain are fewer in number compared to other sea areas. Large-sized fishes such as tuna, bonito, and yellowtails are uncommon. Therefore, middle-and-small sized fishes which become the easy victims of larger predators are plenty. They enjoy a stress-free life without being chased by big fishes here, which definitely affects their flavor in a most positive way.

<What we learn from fishes>

The ocean now experiences rises in temperature and holds less nutrition. The struggle of fishes never ends.

Spawning season now is out of alignment. And natural resources are dying. In such a difficult environment, living things are struggling to maintain their survival. I think they are brave and adorable. Every morning, I see them at the fish market, and they are already fewer in number. And they make me reflect on our behaviors; what we human beings have done to them. And questions arise; what can we do for them? Is there anything we can do for them?

(Contributor, Yuukou Nakamura)

Fish unloaded in Minamata

Tsubodan (Tsubotani)
— Site of Official Discovery and Epicenter —

Map1 D-3

A small mooring area surrounded by rock-lined headland (too small to be a cape) was called Tsubodan. It is also known as Tsubotani, maybe because this area with heights on both sides resembled a valley caused by cliff failure (tani means "valley" in Japanese). Tsubotani also has a tiny shingle (too small to be an inlet), which used to provide a safe playground for children. With no waves or drop-offs, many sea creatures were found under the rocks and pebbles, such as snails, abalones, oysters, small crabs, and sometimes octopuses. A miniature-like dike had a little stone-built shrine and a statue of deity of commerce, which stood over and watched for children. The deity must have sat here quietly observing the entire history of Minamata disease. Two Tanaka girls grew to the age of two and five in the bosom of sea breeze in this tiny playground. Being a family of ship carpenters, the Tanakas' house jutted out over the sea. It looked as if one could enjoy fishing from their window at high tide.

Tsubodan of today (Photographed in 2014)

It was around 1954 when a series of abnormal incidents started to happen to this family. Their cat suddenly dashed into a wall and a pole in the dead of night, jumping and dribbling. It went mad and died soon after. The Tanakas got new cats one after another, but they all went crazy after a few months. Their pig screamed in pain, and became unable to stand up.

Around March 1956, the five-and-11-month-old daughter was found that she could no longer use chopsticks well, and spilled rice. In April, she suffered from unstable walking, slurred speech, insomnia due to crying at night, and dysphagia (difficulty in swallowing). On April 21, she visited the Chisso factory hospital (see p.49). She was admitted to the hospital two days later, but her limbs were almost paralyzed. Every now and then, she was struck by convulsions of entire body. On the day of her sister's hospitalization, the two-and-11-month-old girl complained of pain in her knee and fingers. Her walking was shaky and unstable. She could not grab with her hands nor use chopsticks well. Her speech became unclear. On April 29, the younger girl was hospitalized as well.

Being informed by their mother that the neighbor's girl five and four months old was showing the similar symptoms, the attending pediatrician Dr. Kaneki Noda and the director Dr. Hajime Hosokawa of Chisso factory hospital reported to Minamata Public Health Office (directed by Hasuo Itoh) that "there is an epidemic of an unknown disease of the central nervous system" on May 1. This is the day of official recognition of the disease that we now know as Minamata disease. At first, the disease was suspected of being contagious. Consequently, the patients were transferred to an isolation ward, and their families were ostracized (see p.51). The mother of the Tanaka sisters regretted this until she died; "I asked the doctor if the cat's disease was passed on to my child. I said such a stupid and unnecessary thing, and that's why we got a punishing."

The discrimination was unbearably excruciating for her. Mrs. Tanaka was a plaintiff of the first Minamata disease litigation.

Subsequently, a boy 11 years and eight months old, a boy eight years and seven months old, and his mother successively came down with the disease. The Second Minamata Disease Research Group at Kumamoto University Medical School checked all the 41 residents of eight households in Tsubodan, and diagnosed 25 with Minamata disease and seven with suspected Minamata disease. Only four showed no symptoms.

Even now, people who still live here are suffering painful history of illness and discrimination with patience. This place is literally the epicenter of Minamata disease.

Tsubotani in 1960's (Photograph by Takeshi Shiota)

Yudo
— Area with Many Congenital Minamata Disease Patients —

Map1 C-3

Yudo is one of the areas with the most Minamata disease patients (over 160 were certified). It is particularly marked by a high incidence of congenital Minamata disease.

Fukuro Bay, with its mouth less than 100 m wide, was so calm that only typhoons could agitate it. The bay provided a perfect bed for resting and spawning for migratory fish such as sardine. At the bay mouth, there was a freshwater spring where young mullets used to huddle around.

Yudo in 1877 was a small hamlet with only seven to eight households. Along with the construction of a new fertilizer factory (present Chisso) in Minamata, immigrants from Amakusa settled in the village, which boosted the population. Those immigrants, however, were despised as "islanders", and there were some discrimination and segregation against them even among the fishermen.

Yudo in 1960's (Photograph by Iwao Onitsuka)

In 1953, people in Yudo witnessed unusual behaviors of fish and waterfowls. Soon, cats went mad and died, and finally humans were affected. The repercussion in Yudo was characteristic in that, among mere 30 households along the sea, there were seven congenital Minamata disease patients and nine infantile Minamata disease patients. Ms. Shinobu Sakamoto, a congenital Minamata disease victim who works to eliminate discrimination and advocating a network among people, lives in this district.

Yudo (Photographed in 2006)

Modo

— Area with High Incidents of Minamata Disease where Village Society Was Destroyed —

Map1 B-2

Located at the southernmost part of Kumamoto along the border with Kagoshima, Modo (mo means "lush", and do means "way") is a small fishing hamlet with the highest number of patients suffering from Minamata disease (over 200 have been officially certified). Modo Bay enjoyed relatively calm waters even when the outer sea ran high. In addition, people believed that the shadows of the giant Modo pines along the inlet brought in the fish. Modo was known as an ideal place for fish to escape and spawn, and provided home to mullet, sardine, black sea bream, octopus, squid, and other marine lives. With the rich source of fish and shellfish, the four boat owner families (*amimoto*) and the 120 tenant fisher families (*amiko*) in Modo helped each other and lived like an extended family.

In 1953, most of the domestic cats in the district danced a "convulsion" dance, frothed at their mouths, and dashed into the sea to death. Many people saw black kites and crows fall on the water surface, flap and splash, and die in great agony. The Kumamoto Nichinichi Shimbun newspaper reported on August 1, "Cats Die Off due to Epilepsy in Modo, Minamata, Overwhelmed by Mouse Proliferation". Mr. Torashige Ishimoto, who served as a community leader of the administrative district, urged the City's public health office to eradicate the unbearable upsurge mice. Simultaneously, the abnormal symptoms were appearing on humans as well.

The Modo area typified the devastation of Minamata disease; the lack of fish catch impoverished the fishing hamlet; the Minamata disease victims as well as their family members were discriminated against; and the entire village society (intra-community relationships) was destroyed. But nowadays, there are Minamata disease patients who have been devoted to producing oranges with organic farming method. And the second generations are involving in fishery and growing pesticide-free mandarins.

Modo in 1960's (Photograph by Takeshi Shiota)

Modo (Photographed in 2006)

Otome-zuka（Barrow of Maiden）
― House of Prayer, Exchange, and Learning ―

Map1 A-2

The late Mr. Yoshiharu Tanoue, a Minamata disease patient and plaintiff of the first Minamata disease lawsuit, opened up an agricultural farm in 1978. In 1981, Mr. and Mrs. Akira（deceased）and Emiko Sunada built a barrow at this farm, raising money by performing a monodrama Umi-yo Haha-yo Kodomora-yo（Sea, Mother, and Children）all around Japan. It is a place of worship where all the living things victimized by Minamata disease are honored, and to remember the Minamata disease forever. It was named Otome-zuka（Barrow of Maiden）in memory of Miss Tomoko Kamimura, a fetal Minamata disease patient who died at the age of 21.

The stone chamber at the front of the barrow is covered with clean sand collected from beaches around Japan at the time of fund-raising tour. In it a box made of paulownia is placed that contains memento of Miss Tomoko Kamimura as well as shells from the Jomon-period shell mound in Minamata, papilla moon shells from pre-contaminated Hyakken and

Stone chamber of Otome-zuka

Mategata, bugai from the mouth of effluent-polluted Minamata River, and white round stones from Dragon King Shrine at Osakigahana. It also houses memorials from various places — a-bombed roof tiles from Hiroshima and Nagasaki, corals from Shiraho, Okinawa, a beech stump from the Shiretoko primeval forest, garden stones form the Memorial Hall of the Victims in Nanjing Massacre by Japanese Invaders, and mother-of-pearl shells given by a mother visiting from Micronesia. (from Otomezuka Engi (The History of Barrow of Maiden))

◇ **Annual Events at Otome-zuka**

Memorial Service

The annual memorial service has been held on May 1 every year

Memorial service at Otome-zuka (Photographed in 2006)

by the Minamata Disease Patients Families Mutual Aid Society since 1981, much before the government initiated memorial service which started in 1992, which was the day of official recognition.

◇ Umi-no Boshi-zo (Image of Mother and Child of the Sea)

The original name of the sculpture was Hinshi-no Ko-wo Daku Onna (Image of Woman Embracing a Dying Child), which depicted a mother and a child beleaguered by the typhoon of steel (intensive gunfire) in the battle of Okinawa. The image was renamed as Umi-no Boshi-zo (Image of Mother and Child of the Sea); the dying child represented all the living things, and the mother praying for revival represented the sea.

Image of Mother and Child (bronze, by Minoru Kinjoh, 1972)

Approximately 40 km of the coast between Yatsushiro City to Minamata City is designated as the Ashikita Coast Prefectural Park. The area is known for its deeply indented coastline dotted with fishing villages. Right behind the coastline lies a range of 200 to 300-meter-high mountains, which leaves little land available to cultivate. Therefore, in many terraced fields sweet potatoes and upland rice were grown. Houses were built roof to roof in such narrow and limited area. Access to the district was easier using the sea routes than the passes over the mountains. People in these fishing hamlets lived on a diet of sweet potatoes and fish caught in the sea right in front of them. There is even an old folk tune with the lyrics, "Akasaki and Hirakuni folks don't eat rice. They live on sweet potatoes with a side dish of sardines."

According to the document written by the fishermen in Ashikita Town, purse seine fishing was done in Tsunagi Village in 1936, and soon after the War it was resumed. Purse seines were introduced to the village, and the document states it was great to see numerous fishing boats full of sardines coming back with big-catch flags. It was envy of the Ashikita fishermen.

There are three fishermen's organizations in Fukuura and two organizations in Akasaki, each with a different leader. There are a further five leaders in Egushi, three of whom own a couple of flotillas each. Because of the big-net fishing, they needed 20-30 fishermen who could work under them, and thus fishermen from other areas besides local people were hired.

The ocean-fresh anchovy, a target for purse-net fishing, can suffer damage in seconds. The bosses and fishermen had their own pots to process anchovy. According to the Bulletin of Kumamoto Prefecture Fisheries, there were 344 anchovy-processing factories in Amakusa, 22 in Ashikita, and 40 in Tsunagi. While facilities in Amakusa boasted a large production of dried sardine, Tsunagi had a larger production than either Minamata or Ashikita.

Mr. Tokichi Funaba, who operated a net fishing business along the rocky shores of Tsunagi, became the first certified patient in September 1959. In the same year, a Tsunagi villagers' meeting was held to discuss measures against Minamata disease. A women's society mentioned that having a Minamata disease patient inside the village was a threat, realizing that it would lead to an announcement of a boycott movement and threaten their livelihoods. The picture on p.85 shows that such awareness was not limited to individuals.

As of 2019, there were 354 patients certified by the Pollution-related Health Damage Compensation Law. A further 2,614 people received Minamata Disease Victim's Notebooks. Red-colored houses in the illustration have certified patients who are first to fourth-degree relatives, blue houses have someone who receives a notebook for compensation and relief, and white houses are where the condition for compensation or relief were unknown. A significant majority of the houses are white. This does not mean that the effects of Minamata disease are not widespread. What is more likely is that the area and the families preferred to conceal the disease.

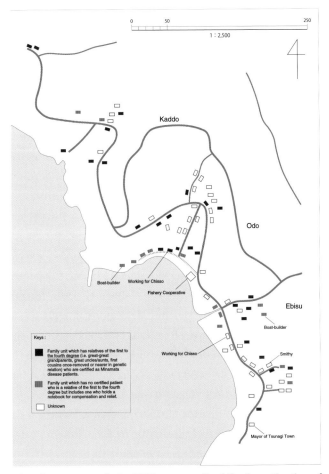

0   50   250
                                    1 : 2,500

Kaddo

Odo

Ebisu

Boat-builder
Working for Chisso
Fishery Cooperative

Boat-builder

Keys :

■ Family unit which has relatives of the first to
the fourth degree (i.e. great-great
grandparents, great uncles/aunts, first
cousins once-removed or nearer in genetic
relation) who are certified as Minamata
disease patients.

▪ Family unit which has no certified patient
who is a relative of the first to the fourth
degree but includes one who holds a
notebook for compensation and relief.

☐ Unknown

Working for Chisso

Smithy

Mayor of Tsunagi Town

Note: the above was made in 2012 as a result of the investigation of each
house by the Open Research Center for Minamata Studies.

Ashikita
— The Coastline Where The Scenery Of The Time Remains —
Map2 B-3

Ashikita Town has some fishing villages including Meshima, Sashiki, Tsurugiyama, and Hakariishi. They are linked through some slaps. The main three of them were named the Santaro Passes: Akamatsutaro, Sashikitaro, Tsunagitaro. These communities had been difficult to access until National Route 3 was constructed. Moreover, many byroads on mountainsides from hamlet to hamlet were hard to pass. The rias coastlines, which are close to the mountains, forced people to go over the mountains even to neighboring villages. As for Meshima community, ship navigation was popular until a prefectural road was constructed in 1963 and cars were thus able to go to the tip of the cape in 1971.

In September 1958, Chisso changed their effluent outlet from Hyakken to the mouth of the Minamata River; and the disease spread northward. The Fresh Fish Retail Association announced a boycott of the fish caught by the fishermen of the Minamata Fishing Cooperative, which had been accepted by many areas. The fishermen were forced into dire poverty. In November 1959, in Meshia, Mr. Fukumatsu Ogata, a leader of fishermen, became the first deceased patient of the disease in Ashikita.

Until one leader of the fishermen and two fetal patients were recognized as Minamata disease patients in 1971, the Meshima Fishing Cooperative, like other fishing cooperatives, had forced its union members not to apply for Minamata disease status. However, in 1972 a fish peddler who had been hospitalized in Izumi City in Kagoshima Prefecture was brought back by the union leaders who tried to conceal the fact, and a liability issue arose between his family and the union. The executive board of the union resolved that the application should depend on a person's individual choice and recommended that everyone who was suffering symptoms should apply for the certification all together. Those who took medical examinations after January in 1973 began to make applications.

As of 2019, there were 346 certified patients by the Pollution-related Health Damage Compensation Law, and 7,259 people received Victim's Notebooks. Inside the dotted line of Meshima (see picture on p. 88), the population is 110 people in 36 families. Among them, 65 residents are certified patients. Families including certified patients who are the first to fourth-degree relatives are colored red. Mapping found all of 36 families red. It indicates that the fishing village, 15 km by air northward from Minamata, was heavily contaminated with mercury, and also that the campaign for uncertified patients, which was active as previously mentioned, had immense influence.

Walking along the coast of the village with such backgrounds, one develops empathy for those sufferers.

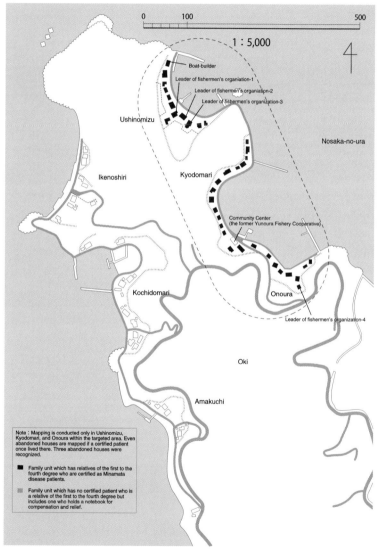

Note: the above was made in 2012 as a result of the investigation of each house by the Open Research Center for Minamata Studies.

Crossing Kaminokawa into Izumi
— Different Prefectures, Same Sea —

Map4 C-2, D-2

Along Route 3 to the south, going 200 meters over a railway underpass in Fukuro, you can see four-meter-wide river, named the Kaminokawa River, also known as the Sakai River. This river marks the border between Kumamoto and Kagoshima Prefectures. The other side of the river is Izumi City in Kagoshima Prefecture, which happens to have a different dialect. Once Satsuma, the former province name including the western part of Kagoshima, prohibited the True Pure Land Buddhism. The believers used to come up to temples in Minamata; they were called 'hidden Buddhists'. Because of this, a checking station named Noma-no-seki along the Satsuma Way conducted strict border administration; however, some people used secret sea routes.

Along the coastline to Izumi, fishing hamlets with high incidents of Minamata disease are nestled: Kizushi, Maeda, Komenotsu Port, Shimosababuchi, Shimochishiki, Sumiyoshi, and Shiomi; further down along the coast are Gata, Shoh, and Warabijima. With the Izumi Plain, the fishing areas here provide somewhat different scenery from the fishing hamlets along the Ashikita Coast or on the islands of Goshonoura. A well-known wintering spot for cranes from Siberia is located there. Over ten thousand cranes can be seen from December to March. Izumi is an old historical town; its name is recognized in the detailed enforcement regulations of the code: *Engishiki* compiled over 1,000 years ago. During World War II it was a base for *kamikaze*; Hiroyuki Agawa, a novelist, wrote about it in his novel titled *Kumo-no-bohyo*. The lead character, a Marine reserve student, did flight training above Minamata before a sortie.

Close to this area lies Katsurajima with a population of eight as of 2018. Heading west, you reach Akune City beside the Kuro-no-seto Strait, whose current is strongest in the Shiranui Sea. A bridge over the strait now connects the island with Nagashima Island. Like Shishijima Island further to the north, the island has a flourishing fishing industry but also many Minamata disease patients.

The first patient in the Izumi area was Mr. Tsurumatsu Kama, a leader among fishermen in Shimosababuchi, who showed symptoms in June 1959 and died 16 months after. The details are depicted in the chapter of what Yuki heard in the book entitled "*Kukaijodo*" written by Michiko Ishimure. Another fisherman in August in Shimosababuchi and a fisherman in September in Shimochishiki developed Minamata disease.

Kagoshima Prefecture has 493 certified patients and over 15,000 people who are subject to the medical project for Minamata disease; most of them are from Izumi.

Nago Fishing Port (Photographed in 2004)

The population of Izumi City in 1960 was 45,214. In the year 2018, out of approximately 54,000 residents, 379 people were certified patients. Nagashima and Shishijima (Nagashima Town) have 83 certified patients. However, the rescue measures are applied to some designated areas, including the former Izumi City before annexation, the former Takaono Town, Azuma-cho in Nagashima, and a part of Akune City. The designation has some convoluted and illogical borderlines.

After the Supreme Court decision on the Minamata Disease Kansai Lawsuit in 2004, the number of applicants for certification abruptly increased. More people started talking about Minamata disease in small steps.

Izumi City is not in Kumamoto Prefecture, but is a neighboring city of Minamata and enjoys a location on the same sea. Utilizing the sea routes, people go back and forth between both cities, and thus no small number of people in fishing villages in Ashikita Town are connected to people in villages in Izumi. As the fish run and migrate freely heedless of the prefectural border, the mercury contamination and its resulting damage spread naturally.

Goshoura-jima Island and Minamata Disease
— Island of 'Hidden' Minamata Disease —

Map3 C-1

Across the Shiranui (Yatsushiro) Sea from Minamata sit two islands: a long, narrow island resembling a sleeping Buddha, and a triangle-like island. They are Goshoura-jima Island and Shishijima Island, respectively. The former is part of Amakusa, Kumamoto Prefecture, whereas the latter belongs to former Azuma-cho in Kagoshima Prefecture. Goshoura Town, which became Goshoura in Amakusa City after annexation in 2006, is 21.57 km^2, is comprised of the three islands of Goshoura-jima, Yokourajima, and Makishima. Main residential districts are Arakuchi, Hongoh, Oura, Motoura, and Kabanoki. Although the town had about 8,500 residents at the time of the Minamata disease outbreak, the population of Goshoura Town had dropped to 2,750 as of 2015. People live primarily on both farming and fishing. In the yard of each house, one may still see variously-sized statues of Ebisu (the Japanese god of fishermen and luck) as well as pots and chimneys for boiling anchovies. The distance from Minamata is approximately 15 km as the crow flies; a small hill on the Island affords a good view of Minamata, and the siren of the Chisso Minamata factory is very audible. The waters of the Minamata River are said to flow directly into Cape Umedonosaki. The legend of Princess Machoga-hime, a lady of great beauty who was descended from the Genji

Oura, Goshoura Town in the 1960s (Photographed by Takeshi Shiota)

clan, says that she was murdered by the remnants of the Heishi clan, and dumped into the sea. Her head was then washed ashore at the cape, where the pine trees sob to mourn the princess. Aside from that tale, the sea current often washes up drowned bodies from off Minamata on the beaches here. During the 1969 Yude River in Minamata City flooding, drowned bodies ended up on this island. It is interesting evidence in examining the fate of the mercury discharged from the Minamata factory.

Once, a patient of Minamata disease made cynical remarks to Dr. Masazumi Harada. "Doc, fish live in the middle of the sea, don't they? Then, how is it possible that the fish caught from the Minamata side cause Minamata disease but the fish caught from the Goshoura side don't?" Up to the investigation in 1971 by the Second Minamata Disease Research Group of the Kumamoto University Medical School, there were officially no patients in Goshoura. It is because the island was never covered by any surveys and nobody applied for certification of Minamata disease. There is no way that no patients existed in Goshoura. In 1959, cats were found to have developed Minamata disease in Shishijima and Goshoura. Over the course of three years from 1962, the public health research laboratory of Kumamoto Prefecture collected hair samples of 2,033 residents of Goshoura to determine the mercury level. The samples of 1,287 people indicate below 50 ppm but over 10 ppm, and 220 samples were over 50 ppm (See the note*). The highest one marked 920 ppm. The woman listed as having 920 ppm mercury in her hair died subsequently without receiving any medical checkups. The person with 357 ppm mercury in the hair was officially certified as a Minamata disease patient after 10 years of great struggle.

What these people suffered from was called "hidden" Minamata disease, named after the hidden Christians after the Shimabara / Amakusa Rebellion in 1637. Though called "hidden", they were in fact shunned. In the early years until the certification of patients was granted, a group of medical students at Kumamoto University and Dr. Shirakawa of Niigata University were investigating to find such patients. Out of over 1,400 applications who had filed, merely 50 odd people were certified as Minamata disease patients. Most of the patients signed to accept the 1996 political settlement. After the decision of the Minamata Disease Kansai Lawsuit, the number of the applicants on this island for certification increased; however, until the issuance of Victim's Notebooks under the Law Concerning Special Measures for the Relief of the Pollution-related Disease, many people recused themselves from applying for relief services to avoid involvement with Minamata disease. Now visiting the island to see the landscape and lives of fishing village will help understand the area's history of Minamata disease. Try private lodging services and feel the local people's connection with the sea.

Additionally, ascending the 442 meter-high Karasu Pass, a 360 degree panoramic view including the Shiranui Sea is offered. The place here is called an island of fossils. Large fish of a previously unknown genus and species from the Cretaceous period were found; the island is a scene of primordial romanticism.

Panoramic view of Goshoura-jima Island

Note*: it is recognized internationally that the 10 ppm level and over of mercury concentration in hair is harmful to humans, and levels between 25 and 50 ppm causes Minamata disease.

Goshoura Quarry

— Island of dinosaurs that has been at the mercy of Minamata disease, and iron or steel slag —

Map3 C-1

Through the environmental pollution prevention project from 1977 to 1990, bottom sludge contaminated with high levels of mercury was dredged, and land was reclaimed from the sea. An impermeable liner was put over the dredged sludge, and also sand from the mountains one meter thick was put on it, and an athletic park named Eco-park Minamata was established there (See p.59). Most of the sand was brought from the Goshonoura Quarry lying to the northwest of Minamata City.

Amakusa's Goshoura-jima Island is a subject for the Remote Islands Development Act. The place name is from a legend which states that a temporary palace for Emperor Keiko was built during his imperial tour. The Goshoura-machi area in Amakusa City consists of 18 small and big islands which include three which are inhabited: Goshonura, Makishima, and Yokourajima. The total size of the inhabited islands is approximately 20 km2 and the population is approximately 2,750 according to the 2015 census data.

Since a group of Kochi University found in 1997 a part of a second thigh fossil, which seemed to be a plant-eating dinosaur's, it became noted as a dinosaur island. In Bentenjima, an uninhabited island, a paw-mark fossil of a meat-eating dinosaur approximately 98,000,000 years ago was found. Goshoura-jima Island has the Amakusa City Goshoura Cretaceous Museum. Opposite Minamata City, on the western side of Goshoura-jima Island, there are three quarries, one of which provided the mountain sand for the previously cited land reclamation at Eco-park Minamata.

After stopping operation, one of the above quarries was reclaimed with iron and steel slag or dredged soil of Yatsushiro Port in 2016. A strongly alkaline pool of water subsequently appeared, and it has been dewatered through neutralization. Thus, people started a residents' campaign, concerned about the influence on the surrounding ocean area.

Recently a plan emerged calling for the mountain sand from one of the above quarries in operation to be taken to Okinawa for Henoko land reclamation.

Pooled water in the remains of a Goshoura Quarry (Photographed in 2016)

How to visit sites in Minamata ~ Model course ~

→walking
⇒bus ride

■ Half-day course

From the ground zero of Minamata disease to the reclaimed land where mercury sludge lies underneath

Minamata station of Hisatsu Orange Railway → Main gate of the Chisso Corporation (currently JNC) Minamata Factory → Hyakken Effluent Outlet / the first site of the Minamata Disease Pilgrimage → Reclaimed land → Shinsui Seawall → *Tamashi-ishi* (Soul Stone) by *Hongan-no-kai* (group for salvation)

Tour of museums

Minamata station of Hisatsu Orange Railway → Minamata Disease Municipal Museum → National Institute for Minamata Disease → Minamata Disease Museum / Soshisha

Hachiman Sedimentation Pool for industrial waste (carbide residue dumping site)

Minamata station of Hisatsu Orange Railway → Chisso's old company housing → Hachiman-Sedimentation Pool → remains of Chisso's Tsukiji apartments / Hachiman company-owned housing (large-scale solar power generation) → Old Chisso factory → Onsite Research Center for Minamata Studies, Kumamoto Gakuen University

Walk in the villages where the Minamata disease outbreak occurred ①

Minamata Station of Hisatsu Orange Railway ⇒ Modo Gyoko (fishing port) bus stop (Minakuru bus) → Modo → Yudo → Tsubodan → Yudo bus stop ⇒ Minamata Station of Hisatsu Orange Railway

Walk in the villages where the Minamata disease outbreak occurred ②

Minamata Station of Hisatsu Orange Railway ⇒ Detsuki bus stop →

Tsubodan → Yudo → Minamata Disease Museum / Soshisha → Jinbara-danchi-mae bus stop (Minakuru bus) ⇒ Minamata Station of Hisatsu Orange Railway

■ One-day course
From the ground zero of Minamata disease to the reclaimed land where mercury sludge lies underneath

Minamata Station of Hisatsu Orange Railway → Main gate of the Chisso Corporation (currently JNC) Minamata Factory → Hyakken Effluent Outlet / the first site of Minamata Disease Pilgrimage → Reclaimed land → Shinsui Seawall → Tamashi-ishi (Soul Stone) by *Hongan-no-kai* (group for salvation) → Minamata Disease Municipal Museum → National Institute for Minamata Disease → Ajino-eki (food station) "Takenko" (lunch) → Minamata Disease Museum / Soshisha

Walk in Chisso-related facilities

Minamata Station of Hisatsu Orange Railway → Main gate of the Chisso Corporation (currently JNC) Minamata Factory → Hyakken Effluent Outlet → Umedo port → Tsuruoka Restaurant (lunch) → Hachiman Sedimentation Pool → Old Chisso factory → Remains of Chisso-affiliated hospital → Minamata Station of Hisatsu Orange Railway

From the ground zero of Minamata disease to the villages where the Minamata disease outbreak occurred

Minamata Station of Hisatsu Orange Railway → Main gate of the Chisso Corporation (currently JNC) Minamata Factory → Hyakken Effluent Outlet / the first site of Minamata Disease Pilgrimage ⇒ Detsuki bus stop → Tsubodan → Yudo → Nanri Restaurant (lunch) → Modo, Modo fishing port ⇒ Jinbara-danchi-mae bus stop → Minamata Disease Museum / Soshisha → Jinbara-danchi-mae bus stop ⇒ Minamata Station of Hisatsu Orange Railway

■ Two-day course

Walk in the fishing villages and isolated islands along Ashikita coast and have a real feeling of mercury pollution and expansion of Minamata disease

The first day: Tsunagi, Akasaki, Hirakuni, Fukuura, Meshima, Sashiki, Tanoura (coastal road)

Take Sanko bus from Minamata Station of the Hisatsu Orange Railway to Tsunagi, Hirakuni, Fukuura, Sashiki or you can access from Tsunagi, Yunoura, Sashiki, Uminoura stations or Higo-Tanoura Station

The second day: Goshoura, Makishima, Shishijima, Izumi (water taxi, share water taxi (refer to p.113)

To isolated islands such as Goshoura, Makishima, Shishijima, take water taxi or ferry boat

To Izumi, take the Hisatsu Orange Railway / Nangoku bus

■ How to get to the destination (access)

Strolling

When time permits, strolling in the town of Minamata, the vicinity of the factory, and the fishing village area is recommended

Reference: It takes 2 minutes from Minamata Station of the Hisatsu Orange Railway to the main gate of Chisso on foot, 10 minutes to the Hyakken Effluent Outlet, 15 minutes to San-no jinja, 30 minutes to the Shinsui Seawall and 90 minutes to Modo

Renting a bicycle

On a sunny day, exploring the town by bicycle while enjoying a comfortable breeze is recommended

In the city of Minamata, "developing a town for bicycles" for people and the environment has been promoted. Using a bicycle share system dubbed "Charizo-kun", visitors can rent bicycles for free upon receiving a temporary ticket for rental bicycles.

Reference: Rental stations will be at Michino-Eki Minamata, Shin-Minamata Station, Minamata Station and M's City

Regional Promotion Group, General Affairs Section, Minamata City (0966-61-1607)
Lending situation https://f-cs.jp/sf-r/?id=mina0001

Taking public transportation (bus)

Public transportation (Sanko Bus, Nangoku Bus and Minamata City Community Bus (Minakuru Bus) is available to access various areas in the city (Fukuro, Modo, Yudo, Minamata-ko (port) and others), Yunotsuru Onsen, Yunoko Onsen, the Izumi / Akune area, Kagoshima Airport and others.

Contact information: Sanko Bus (0966-63-2185),
Nangoku Bus (0966-62-1626),
Regional Strategy Office, Planning Section, Minamata City (0966-61-1607)

Renting a car

Renting a car is one way for a group of 4 to 5 people.

Contact information: Minamata Rent-a-Car (Umedo-machi) 0966-53-2727,
Mo-mo Rent-a-Car (Hatsuno) 0966-63-5888,
Tobu Rent-a-Car (Nakazuru) 0966-62-7001

Taking a water taxi

Water taxi : 11 companies at Hongo port, 12 at Arakuchi port, 2 at Oura port, 2 at Yokoura port, 6 at Yoichigaura port and many more.

The fare is approximately 10,000 yen for one way (reservation required).

Contact information: Tourist Information Center in the Goshoura local product shop called "Shiosai-kan" 0966-67-1234

Place to Study and Convey the Lessons from Minamata Disease

Minamata Disease Center Soshisha

Map1 C-3

This general incorporated foundation was established to conduct research on Minamata disease and public awareness activities of the disease. It also provides consultation for Minamata disease patients and those involved, helping them solve problems related to their daily lives. The activities include offering tours of and information on Minamata, lectures, publishing of the newsletter *Gonzui,* maintaining its website, and collecting, organizing, providing materials, and disseminating information.

34 Fukuro, Minamata, Kumamoto 867-0034

Phone: 0966-63-5800   Fax: 0966-63-5808

http://www.soshisha.org/jp/   Email: info@soshisha/org

Minamata Hotaru-no-Ie (Tomi-no-Ie)

Map1 D-5, Map1 C-2

In 1996, when the Minamata disease case was said to be officially and completely settled, Hotaru-no-Ie was established to address the ongoing issues of Minamata disease with patients and sufferers as a place representing a small lit torch like a *hotaru,* a firefly. It has kept working as a secretariat of sufferers, helping them apply for official certifications, organizing medical checkups, and giving general and medical counseling.

Minamata Hotaru-no-Ie

108 Uchiyama, Nampukuji, Minamata, Kumamoto 867-0023
Phone/Fax: 0966-63-8779 (Prior notice is essential before visiting)
Minamata Disease Collaboration Center, Tomi-no-Ie (non-profit organization)
705-4 Fukuro, Minamata, Kumamoto 867-0034
Phone: 0955-62-0280

Sakae-no-mori Hottohausu
Map1 F-5

This social welfare corporation was established starting with a program of giving patients' talks to elementary schools in November 1988. Now our targets have expanded to students of secondary or higher educations and researchers. The members strive to convey the reality of Minamata disease as a spoken fabric of collaboration, with warp of words by patients and woof of messages by those who give them cuddles. Hottohausu is named after "hotto suru" which means "to be relieved"; it tries to be a haven of healing for sufferers of Minamata disease and others in difficulties. The members regard it as their responsibility and duty to pass on the truth of Minamata disease.

1-9-17 Hama-machi, Minamata, Kumamoto 867-0065
Phone: 0966-62-8080   Fax: 0966-83-6200
http://hottohausu.blog18.fc2.com
Email: hottohausu@mx35.tiki.ne.jp

Joint Enterprise Cooperative Econet Minamata
Map1 C-3

For over forty years, this joint enterprise cooperative has made efforts to produce ecological soaps as well as pesticide and chemical fertilizer-free mandarin oranges. Learning from past experience that nature and humans cannot be restored to what they used to be once the environment and nature are destroyed. Reflecting on the way of life of mass production/

consumption/disposal, it offers ideas to lead to a better future life and safe/secure products unique to Minamata. In addition to a rental hall, walking tours around Minamata and information regarding the current situation are also available.

In January 2005, the Minamata Soap Factory and Minamata *Hannoren* (the anti-pesticide union of Minamata) were merged with each other, and established the Joint Enterprise Cooperative Econet Minamata with seven capital providers.

60 Nanpukuji, Minamata, Kumamoto 867-0023

Phone: 0966-63-5408   Fax: 0966-63-3522

http://www.econet-minamata.com

Email: info@econet-minamata.com

Gaia Minamata

Map1 C-4

It is a small organization of five members. They used to be fishermen but are now farmers producing Amanatsu oranges. They have helped each other through Minamata disease incidents since 1977. Although their main job is selling and processing the oranges, they have been engaged in social movements such as supporting the Mizoguchi dismissal revocation suit against the government.

1-39 Fukuro, Minamata, Kumamoto 867-0034

Phone: 0966-62-0810   Fax: 0966-62-0814

http://gaia.iinaa.net

The Minamata Environmental Academia

Map1 E-4

Providing opportunities to learn policies of environmental management and public administration, it works as a hub to offer information on the technical development and dissemination of environmental measures. It is managed by Minamata City and others, and works as a measure for regional development based on the Law Concerning Special Measures for the Relief of Minamata Disease Victims.

6-1 Nanpukuji, Minamata, Kumamoto

Phone: 0966-84-9711   Fax: 0966-84-9713

https://minamata-web.jp/academia/MyHp/Pub/

Email: academia@city.minamata.lg.jp

Open: 9:00 - 19:00   Closed on Sundays, Saturdays, and holidays

NPO Minamata

Map1 E-4

It was established in the wake of the out-of-court settlement of the third Minamata disease litigation. It promotes management of the materials and data related to Minamata disease such as litigation documents and dissemination of the results of research.

2-2-20 Sakurai-machi, Minamata, Kumamoto 867-0045

Phone: 0966-62-9822   Fax: 0966-62-1154

http://minamata.org/

Email: npo@minamata.org

Association of Transmitting Stories of Minamata Disease: *Minamata byo wo kataritsugukai*

Map1 E-2

Led by storyteller Ms. Rimiko Yoshinaga, this general incorporated association was established in 2013 to pass on knowledge of Minamata disease to younger generations. It promotes activities such as offering study meetings of Minamata disease, inviting lecturers, and cultivating human resources as guides for the Minamata Disease Municipal Museum.

1-14 Myojin-cho, Minamata, Kumamoto 867-0055

Phone: 0966-83-7181   Fax: 0966-62-1799

http://www.fbwminamata.com

Email: fbwminamata@gmail.com

Amakusa City Isana-kan

Map3 C-1

3527 Goshoura, Goshoura-machi, Amakusa, Kumamoto 866-0325

Goshoura Branch Office of Amakusa City 2 F

Moyai-kan, Minamata City Moyainaoshi Center

Map 1 F-5

3-1 Makinouchi, Minamata, Kumamoto 867-0005

Phone: 0966-62-3120   Fax: 0966-62-3130

Orange-kan, Minamata City Nanbu Moyainaoshi Center

Map 1 D-3

195-2 Tsukinoura, Minamata, Kumamoto 867-0035

Phone/Fax: 0966-62-2111

Kizuna-no-sato, Ashikita Town Moyainaoshi Center, Ashikita Public Health Center

Map2 C-4

1439-1 Yunoura, Ashikita-machi, Ashikita-gun, Kumamoto 869-5563

Phone: 0966-86-0349

Fax: 0966-86-0410

Yumemoyai, Ashikita Town Meshima Katsudo-suishin (renovation) Center "Yumemoyai"

Map2 B-3

770-14 Meshima, Ashikita-machi, Ashikita-gun, Kumamoto 869-5564

Facilities to Learn about the Minamata Disease Incident

Minamata Disease Municipal Museum

Map1 E-2

Minamata disease is the starting point of pollution-related disease, and its memory should never fade with time. The museum was established to collect and save the invaluable documents and materials of Minamata disease in order to pass them on to the next generations. Introducing the miseries of pollution by photographs and videos, it has employed oral exhibition by patients about their painful firsthand experiences of discrimination.

Approximately 50,000 people now visit the museum every year, which serves as a place of education on the topics of pollution and environment as well as human rights. The displays were redesigned in 2016.

53 Myojin-cho, Minamata, Kumamoto 867-0055

Phone: 0966-62-2621   Fax: 0966-62-2271

http://www.minamata195651.jp

Open: 9:00 - 17:00

Admission: Free

Closed on Mondays (Tuesday if Monday falls on a national holiday or a substitute public holiday) and year-end through the New Year holidays (December 29 - January 3)

Minamata Disease Museum of Minamata Disease Center Soshisha

Map1 C-3

It was established to remember long in our hearts the Minamata disease incident caused by Chisso. The exhibition is divided into four sections with

four themes: The Shiranui Sea, the bountiful sea and life; Minamata disease, Chisso's crime; Struggle, Sufferers' Paths; Present, What we should do. Exhibits under these themes are composed of approximately 100 photos and expository panels, a hut where the cat experiment was conducted, wooden boats, fishing gear, products of Chisso, and others. It points out that reviewing your lifestyle is also essential while questioning the modern society that brought about Minamata disease.

34 Fukuro, Minamata, Kumamoto 867-0034

Phone: 0966-63-5800   Fax: 0966-63-5808

http://www.soshisha.org/jp/koshokan

Open: 9:00 - 17:00

Admission (tax not included) : ¥500 for adults, ¥400 for high school students, ¥300 for junior high school and elementary school students

Admission free for residents of Minamata City, Ashikita Town, Tsunagi Town, Izumi City, Goshourajima Island, and Shishijima Island.

Closed on Saturdays and year-end through the New Year holidays (December 28 - January 4)

Onsite Research Center for Minamata Studies, Kumamoto Gakuen University

Map1 F-5

2-7-13 Hama-machi, Minamata, Kumamoto 867-0065

Phone: 0966-63-5030

FAX: 0966-83-8883

http://www3.kumagaku.ac.jp/minamata/

Email: m-genchi@kumagaku.ac.jp

Admission: Free

Open: 10:00 - 16:00

Closed on Saturdays, Sundays, Mondays, national holidays, year-end through the New Year holidays and others; please contact us.

National Institute for Minamata Disease

Map1 G-5

It was established in October of 1978 in Minamata City, Kumamoto Prefecture, to improve medical treatment for the Minamata disease patients. Having balanced consideration of Minamata disease as the origin of pollution which occurred in Japan, its profound historical background and social importance, it conducts comprehensive medical research to promote better measures against Minamata disease. In July 1996, the institute underwent a structural reorganization to form the new National Institute for Minamata Disease.

4058-18 Hama, Minamata, Kumamoto 867-0008

Phone: 0966-63-3111   Fax: 0966-61-1145

http://www.nimd.go.jp/

NIMD Minamata Disease Archives

Map1 E-2

It was established in June 2001 to widely offer information relating to Minamata disease by collecting, organizing, and storing such data and materials in an integrated fashion, and to contribute to research activities on Minamata disease. It is equipped with an exhibition room, document rooms, an auditorium, a Minamata disease consulting room, a social science research room, and an environmental science laboratory. The exhibition room and document rooms are open to public.

55-10 Myojin-cho, Minamata, Kumamoto 867-0055

Phone: 0966-69-2400   Fax: 0966-62-8010

http://www.nimd.go.jp/archives/index.html

Open: 9:00 - 17:00

Closed on Mondays (Tuesday if Monday falls on a national holiday or a substitute public holiday), and year-end through the New Year holidays (December 29 - January 3)

Kumamoto Environmental Education and Intelligence Center

Map1 E-2

The center is located on scenic Cape Myojinzaki, at the westernmost tip of Minamata City overlooking the Yatsushiro Sea. Regarding Minamata disease as a lesson, it focuses on various global and local environments and how to coexist and live in harmony with nature, a water contamination simulation, vehicle exhaust experiments, hands-on experience of waste-free shopping and sorting waste, research of carbon dioxide absorption by trees and lectures. Through these activities, visitors can learn eco-friendly lifestyles.

55-1 Myojin-cho, Minamata, Kumamoto 867-0055

Phone: 0966-62-2000　Fax: 0966-62-1212

http://www.kumamoto-eco.jp/center/

Open: 9:00 - 17:00

Closed on Mondays (Tuesday if Monday falls on a national holiday or a substitute public holiday) and year-end through the New Year holidays (December 29 - January 3)

International Mercury Laboratory

Map1 C-3

It was established as a hub for the following three aims: 1. to proceed further with the studies on mercury dynamics in vivo and in ecosystem, and on high-sensitive and high-precision mercury analysis technology for environmental and biological monitoring though they have been developed and refined so far; 2. to conduct technological development research on mercury removal and recovery from mercury-containing wastes and other materials; and 3. to offer international technological cooperation and assistance in environmental fields against increasing mercury contamination abroad through the above research achievements.

426-2 Fukuro, Minamata, Kumamoto 867-0034

Phone/Fax: 0966-63-0810

Contact Information for On-site Tours

The Minamata-Ashikita Pollution Research Circle

It was established in 1976 by mainly local elementary, junior high, and high school teachers. Since then, the members have taken care of on-site tours of Minamata for students and teachers from in and out of Kumamoto Prefecture. In addition, they give lectures to school teachers and office staff. Concerned that Minamata disease has been invisible, they would like people put their feet in the patients' shoes not only to recognize Minamata disease as a piece of knowledge. They try to make opportunities to talk with patients in person.

Contact person: Takuji Umeda

2-9-7 Yamate-cho, Minamata, Kumamoto 867-0046

Phone: 0966-63-8516

Econet Minamata

Map1 C-3

In 1987, learning from the lesson of Minamata disease, in cooperation with 54 members including Minamata disease patients, the Chisso workers, citizens in Minamata, a factory named the Minamata Soap Factory was established and reorganized into a joint enterprise cooperative named Econet Minamata in 2005. Now it promotes an ecological movement. As for the on-site tour guides, they are patients, the former Chisso workers, and civic activists. They convey Minamata in its entirety as it really is while talking about occurrences during the Minamata disease and the current situation regarding the Minamata disease incident, which is not over yet. The tour fee is negotiable.

60 Nanpukuji, Minamata, Kumamoto 867-0023

Phone: 0966-63-5408   Fax: 0966-63-3522

http://www.econet-minamata.com

Email: info@econet-minamata.com

Kan-Shiranui Planning

Map1 E-4

This general incorporated association offers coordination and organization of educational tours and observation-study on site in Minamata and Ashikita for visitors inside and outside of Kumamoto. Recommended are the programs focused on fieldwork by Minamata citizens or experience of creating or cooking things by making the most of nature. It also promotes some community planning projects which include developing specialty products and creating new tourism utilizing local resources: bountiful nature of mountains and the sea in Minamata and Ashikita.

Nishida building 1F, 2-4-8, Showa-machi, Minamata, Kumamoto 867-0051

Phone: 0966-68-9450   Fax: 050-3730-3585

https://www.kanpla.jp/

Email: info@mkplan.org

Minamata Disease Center Soshisha

Map1 C-3

34 Fukuro, Minamata, Kumamoto 867-0034

Phone: 0966-63-5800

Fax: 0966-63-5808

http://www.soshisha.org/jp/

（See p.106: Minamata Disease Museum）

Accommodation and hot springs in Minamata

▮ Hotels and inns around Minamata Station

Super Hotel Minamata
1-1-38 Daikoku-cho, Minamata-shi
Phone: 0966-63-9000
http://www.superhotel.co.jp/h_links/
minamata/minamata.html
minamata@superhotel.co.jp

Business Hotel Sunlight
3-2-31 Sakurai-cho, Minamata-shi
Phone: 0966-63-0045
sunlight@axel.ocn.ne.jp

▮ Yunoko Hot Spring

Map1 H-6
15-minute bus ride or 1-hour walk
from Minamata Station

Saito Ryokan (lodging and day-visit spa)
Phone: 0966-63-2463

Shoyokan (lodging and day-visit spa)
Phone: 0966-63-4121
http://www.shoyokan.com
info@shoyokan.com

Hiranoya Ryokan (lodging without meals and day-visit spa)
Phone: 0966-63-2161
info@hiranoya.jp

Yunoko Umi-to-Yuyake (lodging and day-visit spa)
TEL 0966-62-6262
http://www.umitoyuyake.com/
info@umitoyuyake.com

Matsubara-so (lodging and day-visit spa)
Phone: 0966-63-2723
http://www.matubaraso.jp/
yunoko@matubaraso.jp

Yuhi-no-Yado (lodging and day-visit spa)
Phone: 0966-83-9727

Yunoko-so (lodging)
Phone: 0966-63-3591

Onsen Iwasaki (lodging and day-visit spa)

Phone: 0966-62-3354
http://onsen-iwasaki.sakura.ne.jp

Shiraume Yurara（lodging and day-visit spa）
Phone: 0966-63-3151
http://www.kaigocsc.co.jp/shiraume/yurara

■ Yunotsuru Hot Spring

Wide-area map F-4
20-minute bus ride or 2-hour walk from Minamata Station

Asahi-so（lodging and day-visit spa）
Phone: 0966-68-0111
https://www.asahisou.jp/
info@asahisou.jp

Kikuya Ryokan（lodging and day-visit spa）
Phone: 0966-68-0211

Tsurumi-so（lodging and day-visit spa）
TEL 0966-68-0033

http://turumisou.com/

Tojiya（lodging and day-visit spa）
Phone: 0966-68-0008
http://www.tojiya.net

Hotaru-no-Yu, Yunotsuru Onsen Health Center（day-visit spa）
Phone: 0966-68-0811

■ Goshoura

Map3 C-1
30 minutes by water taxi from Minamata Port
Regular liner operation: three services every day.
Booking passage on a taxi in the morning is needed by 17:00 the day before the voyage, and in the afternoon needed by one hour before the voyage.
For booking, call Shiosai-kan at 0969-67-1234.
Accommodation is available at one or two dozen inns, hotels, or guest houses. For details, please contact the tourist information center at **Shiosai-kan, Goshoura Bussankan**（product promotion center）.
Phone: 0969-67-1234
Goshoura.net   http://www.goshoura.net/

Where to eat, have a break, and shop

Note: In addition to the following, there are many other restaurants in the downtown area.

▇ Minamata City

Aji-no-Eki Takenko
Map1 D-3
Phone: 0966-63-5501 (champon, lunch buffet)

Yunoko Spain Village, Fukuda Farm
Map1 G-6
Phone: 0966-63-3900 (paella)

Kaijiru-ajidokoro Nanri
Map1 C-3
Phone: 0966-62-4200 (clam soup, cutlassfish on rice)

Fukinotou Artisanal Bakery
Map1 E-4
Phone: 0966-62-2910 (freshly-baked bread)

Suisen Shokudo
Map1 E-4
Phone: 0966-63-6123 (pub, champon)

Napoleon
Map1 F-5
Phone: 0966-63-2328 (Western cuisine)

Isshintaisho
Map1 F-5
Phone: 0120-601-132 (pub, deep-fried banded blue-sprat)

Mont-Vert Nouyama
Phone: 0966-68-0028 (broiled porc)

Samukawa-suigen-tei
Phone: 0966-69-0776 (Somen-nagashi in summer)

Montblanc Fujiya
Map1 E-4
Phone: 0966-63-1179 (Western confectionery)

Ganzo
Map1 E-4
Phone: 0966-63-2136 (pub, fish cuisine)

Mojocado
Map1 E-4
Phone: 0966-83-5004 (processed agricultural foods and souvenirs of Minamata)

Kiraku-shokudo
Map1 F-5
Phone: 0966-62-2629（set meals, champon）

Amando
Map1 E-4
Phone: 0966-62-5020（café）

Yanagiya-honpo
Map1 E-5
Phone: 0966-63-2239（Miki-monaka: wafer cake filled with bean jam）

Tsuruoka Shokudo
Map1 F-3
Phone: 0966-62-3436（champon noodle）

Minamata City Fureai Center (Machikado Rest Lounge)
Map1 E-4
Phone: 0966-84-9909（Minamata tea; carry-in drinks are allowed）

Santaromochi-honpo
Map1 F-4
Phone: 0966-62-2669（Santaro-mochi: rice cake）

Sky Restaurant M's
Map1 F-4
Phone: 0966-63-3252（buffet）

Shokudokoro Kashiwagi
Phone: 0966-68-0031（eel dishes, March - October, reservation needed）

Tsurunoya, Yunotsuru Geihinkan
Phone: 0966-68-0268（lunch buffet）
http://tsuru-noya.com/

Shokokuya-honpo
Phone: 0966-68-0001（café of old private house, lunch）

Tana Cafe, Airinkan（Minamata City Kugino Furusato Center）
Map2 D-6
Phone: 0966-69-0485（café, Thai curry）

Arlequin Confectionery-Atelier
Map1 E-4
Phone: 0966-63-7558（Western confectionery）

Horaku Manju
Map1 E-4
Phone: 0966-63-2673（kaitenyaki: pancake stuffed with bean-jam, shaved ice）

Sogi-no-Taki
Map1 E-4
Phone: 0966-63-2340（set meals, fish dishes）

Yajiro Shokudo

Phone: 0966-62-2846 (set meals, champon)

▓ Ashikita Town

Michi-no-Eki Tanoura

Map2 B-2

Phone: 0966-87-2230 (cutlassfish sushi)

Michi-no-Eki Ashikita Decopon

Map2 C-3

Phone: 0966-61-3020 (Ashikita beef and Decopon)

▓ Goshoura Town

Shoen

Map3 C-1

Phone: 0969-67-2427 (fish dish)

Sushimasa

Map3 C-1

Phone: 0969-67-2405 (brawn cuisine, reservation two days prior to arrival required)

A Brief Chronology of the Minamata Disease Incident

Date	Event
August 1908	Nippon Chisso Hiryo K.K. (Japan Nitrogen Fertilizer Company) established in Minamata.
April 1926	Chisso pays lump-sum solatium of 1,500 yen to the Minamata fishery cooperative for damage incurred by local fisheries because of Chisso's wastewater.
May 1932	Chisso Minamata factory begins production of acetaldehyde, draining its industrial wastewater into Hyakken port.
December 1953	A girl of Detsuki, Minamata City, develops unknown neurological disease (later confirmed as the first Minamata disease patient).
June 1954	Most cats in Modo, Minamata, die mad.
April 1956	A girl of Tsukinoura, Minamata City, consults Chisso Factory Hospital. (21st)
May 1956	Chisso Factory Hospital reports to Minamata Public Health Office about "four patients suffering unknown disease" (later recognized as the official confirmation of Minamata disease). (1st)
April 1957	A cat develops Minamata disease after being fed fish from Minamata Bay in experiments by Hasuo Ito, director of Minamata Public Health Office.
September 1957	Ministry of Health and Welfare decides not to apply the Food Sanitation Act that bans the catch and sales of contaminated fish.
September 1958	Chisso reroutes its acetaldehyde effluent discharge from Hyakken drain outlet to Hachiman Sedimentation Pool, thus draining its waste into Minamata River mouth.
July 1959	Kumamoto University research group announces "Minamata disease is a neurological disorder caused by consuming fish and shellfish from Minamata Bay, and

mercury has drawn a great deal of attention as the polluting toxin."

October 1959   Hajime Hosokawa, director of Chisso factory hospital, confirms through his cat experiment that feeding of factory wastewater causes Minamata disease.

November 1959   Rally calling for cessation of factory operation organized by over 2,000 Shiranui Sea inshore fishers, who stormed into Chisso factory (what is called as Fishers' Rebellion). (2nd)

November 1959   Minamata Disease Food Poisoning Survey Team of the Food and Sanitation Investigation Council of Ministry of Health and Welfare reports"Minamata disease is a toxic disorder that chiefly affects the central nerve system, caused by a large consumption of fish and shellfish living in and around Minamata Bay, and mainly attributable to some kind of organic mercury compound."

November 1959   Sit-in in front of Chisso Minamata factory staged by Minamata Disease Patients' Families Mutual Aid Society, demanding a flat sum of ¥3,000,000 per patient for compensation.

December 1959   Installation of effluent treatment facility (Cyclator) completed at Chisso Minamata factory.

December 1959   Patients' Families Mutual Aid Society concludes "sympathy money contract" with Chisso: ¥300,000 for the deceased, annual payments of   ¥100,000 for adults and ¥30,000 for children; and further compensation claims relinquished even if it is decided in the future that Minamata disease is caused by Chisso. (30th)

April 1962   Chisso proposes the stable wage system to its labor union, which provoked the employer-employee dispute including indefinite strike.

November 1962   The Certification Committee officially recognizes 16 patients with fetal or congenital Minamata disease. (29th)

May 1965	Professors Tsubaki and Ueno of Niigata University report to the public health department of Niigata Prefecture on"sporadic outbreak of mercury-poisoned patients of unknown cause in the area along lower Agano River". Official recognition of Niigata Minamata disease outbreak.
June 1967	13 Niigata Minamata disease patients of three households sued Showa Denko in the Niigata District Court.
May 1968	Chisso stops producing acetaldehyde by the acetylene process.
August 1968	Chisso Labor Union makes a statement"it is shameful to have done nothing for the battle against Minamata disease to this day", known as the Shame Resolution.
September 1968	The national government designates Minamata disease as a pollution-triggered disease, announcing"Kumamoto Minamata disease was caused by methyl mercury compounds produced in the acetaldehyde acetic acid facilities of the Chisso Minamata Factory." (26th)
June 1969	112 patients (in 29 households) of the Trial Group of Patients' Families Mutual Aid Society bring a civil suit against Chisso with the Kumamoto District Court asking for compensation of total ￥640,000,000 (what is referred to as"the first lawsuit"). (14th)
December 1969	The committee for examining the designation of pollution-triggered diseases (Ministry of Health and Welfare) officially names the mercury poisoning disease as "Minamata disease". (17th)
August 1971	In response to the complaint made by Teruo Kawamoto and other patients under the Administrative Complaint Appeal Law, the Environment Agency overturns the original judgment of Kumamoto Prefecture that dismissed the application for certification. Vice minister announces patients' certification.
October 1971	Newly certified patients such as Teruo Kawamoto starts

	the direct negotiations with Chisso (so-called direct negotiation fight). Sit-ins in front of the main gate of Minamata factory and the Tokyo headquarter of Chisso staged for 21 months.
June 1972	Delegates including Tsuginori Hamamoto and Shinobu Sakamoto participate in the UN Conference on the Human Environment (at Stockholm), publicizing the reality of Minamata disease to the world.
March 1973	The first lawsuit ruled in favor of plaintiffs. Minamata disease negotiating team starts direct negotiations with Chisso.
July 1973	Compensation agreement between Minamata disease patients and Chisso reached.
August 1974	Minamata Disease Certification Applicants' Council established.
March 1975	Minamata disease patients' group at Kansai area forms Mutual Aid Society for Minamata Disease Patients' Families in Tokai Area.
December 1976	Verdict of a case asking for confirmation of illegality of administrative inaction rendered in favor of plaintiffs, ruling that the delay of certification process was violation of law.
July 1977	Environment Agency issues a notice entitled"On Criteria for Diagnosis of Acquired Minamata Disease" in the name of the Director of the Environmental Health Department.
March 1979	Kumamoto District Court convicts the former president of Chisso (guilty verdict finalized at Supreme Court in February 1988).
May 1980	The third Minamata disease suit filed, demanding national compensation for the first time.
October 1982	The Kansai Minamata disease suit filed, demanding national compensation.
July 1988	Patients applying for certification file for the cause-effect

adjudication on Minamata disease with the Environmental Dispute Coordination Commission, which refused to accept the application in September.

September 1988   Minamata disease negotiating team stages a sit-in in front of Chisso Minamata factory, demanding acknowledgement of patients as victims and redress for them.

May 1994   Masazumi Yoshii, Mayor of Minamata City apologizes for the first time as the head of the municipality at the Minamata disease memorial ceremony for the victims.

July 1994   Osaka District Court hands down its decision on the Minamata disease Kansai case, ruling out the liability of the state and the prefectural governments. Plaintiffs appeal.

October 1995   Five uncertified patients' groups accept the final political settlement proposal by the national government.

April 1996   Minamata Disease Patients Alliance and Chisso conclude the agreement.

May 1996   National Liaison Conference of Plaintiffs Counsels for Minamata Disease Victims and Chisso conclude the agreement.

August 1997   Removal of the Minamata Bay containment net launched (completed in October).

June 1999   Cabinet approves a drastic public bailout plan for Chisso.

April 2001   Osaka High Court finds the national and prefectural governments responsible on the Minamata disease Kansai case. The governments appeal.

September 2002   Kumamoto Gakuen University starts a series of lectures on"Minamata studies" (which has since been offered annually).

October 2004   Supreme Court makes its judgment on Minamata disease Kansai case, recognizing the liability of the state and the local governments. The decision becomes final and binding.

April 2005	Kumamoto Gakuen University establishes the Open Research Center for Minamata Studies.
August 2005	Kumamoto Gakuen University opens Onsite Research Center for Minamata studies in Minamata City.
October 2005	Shiranui Group of Minamata Disease Patients filed a damage suit against the state, Kumamoto Prefecture, and the offending enterprise Chisso.
September 2006	The private consulting group of Minamata disease issues, led by Minister of the Environment Koike, does not revise the standard of certification for Minamata disease.
	The International Forum on Environmental Pollution and Social Impact by Kumamoto Gakuen University is held.
March 2007	Masami Ogata (49), filing an administrative complaint, is certified as a Minamata disease patient.
April 2007	The third Niigata Minamata disease lawsuit at the Niigata District Court.
May 2007	The Kawakami mandamus suit for certification in the Kumamoto District Court and F's lawsuits in the Osaka District Court are filed. They demand to nullify the dismissal of certification.
October 2007	The Mutual Aid Society for Minamata Disease Patients files a suit for damages against the national and the prefectural governments and Chisso.
January 2008	The Kumamoto District Court ruling on the Mizoguchi mandamus lawsuit to demand to nullify the dismissal of certification is against the plaintiff.
February 2008	The Mizoguchi lawsuit is appealed at the Fukuoka High Court.
December 2008	A 58-year-old man in Kagoshima is certified as a Minamata disease patient for the first time in eight years. The number of the applicants for the certification exceeds 6,000.

July 2009	The Act on Special Measures, which accepts the company split-up of Chisso and measures for the relief of the uncertified patients, passes.
June 2010	A junior high school student is told by some students of another city during a soccer practice match, "Minamata disease! Don't touch me."
July 2010	At the suit to nullify the dismissal of certification and the mandamus action for certification, the Osaka High Court denies the current certifying standard because of no medical justification, and recognizes the plaintiff as a Minamata disease patient. The Kumamoto prefectural government appeals the decision.
January 2011	Chisso establishes a subcompany named JNC Corp. to take over Chisso's business division.
March 2011	The group lawsuits by the Shiranui Group of Minamata Disease Patients are settled at three district courts.
February 2012	The Mizoguchi lawsuit wins a reversal.
June 2012	Dr. Masazumi Harada passes away.
July 2012	The application based on the Special Measures to rescue uncertified Minamata disease patients is closed. The total number: 65,151 applicants is over twice as many as expected.
February 2013	The Kagoshima and Kumamoto prefectural governments indicate the intention to deny the motion by those who are excluded from the Special Measures Law.
March 2013	The Niigata prefectural government indicates the intention to accept the appeal
April 2013	For the Mizoguchi lawsuit, the Supreme Court recognizes her as a Minamata disease patient for the first time.
June 2013	Forty-eight people who are not covered by the Special Measures Law file a suit for damages against the national government, prefectural governments, and Chisso.
October 2013	The Minamata Convention on Mercury to protect human

health and the environment from pollution is adopted.

October 2013   The Imperial couple visits Minamata City for the first time for the National Convention for the Development of Abundantly Productive Sea, and pays a visit to patients of Minamata disease.

October 2013   The board of administrative appeal decides in the Shimoda's case to nullify the prefectural decision and rules that Shimoda merits certification.

March 2014   The Ministry of Environment notifies the guidelines regarding the management for the standard of the Minamata disease certification. The ministry does not change the standard, but ask the sufferers for some objective evidence such as that of mercury exposure. They react sharply against it.

March 2014   At the lawsuit by the Mutual Aid Society for Minamata Disease Patients, the Kumamoto District Court orders compensation for three out of eight uncertified patients who were plaintiffs.

April 2014   The Kumamoto prefectural government returns the operation of screening for the certification to the national government, so the Special Certification Council for Minamata Disease resumes for the first time in twelve years.

May 2014   The president of the Mutual Aid Society for Minamata Disease Patients, Sato, files a mandamus suit to research the damage of Minamata disease according to the Food Sanitation Law by the national government and Kumamoto Prefecture.

June 2014   The law related the Companies Act, which allows Chisso to sell its subcompany without the approval at the stockholder meeting, is passed at the Upper House plenary session on 20 June.

August 2014   The outcome of the relief measure based on the Special

Measures Law for those who apply for a lump-sum payment by the Kumamoto, Kagoshima, and Niigata prefectural governments is as follows: 19,306 people in Kumamoto, 11,127 in Kagoshima, and 1,811 in Niigata (as of 22 June) for a lump-sum payment; 6,013 only for a Minamata Victim's Notebook; and 9,649 applicants are excluded.

December 2014   The Kagoshima prefectural government certified a 60-year-old man in Izumi City as a Minamata disease patient. (The certified patients in Kagoshima Prefecture: 493 people.)

June 2015   The marbled rockfish with more than double the provisional regulatory concentration of mercury 0.4 ppm around the Minamata Port are recognized.

October 2015   Seven members of the Mutual Aid Society for Minamata Disease Patients file a suit with the Kumamoto District Court for the mandamus action for the certification as Minamata disease patients based on the Pollution-related Health Damage Compensation Law.

November 2015   The Kumamoto prefectural government certified a 60-something woman as a Minamata disease patient for the first time after the notice of the new guidelines regarding the management of the certification standard (in March 2014).

December 2015   Litigation against the Minamata disease certification standard new notice, plaintiff defeated in Supreme Court.

February 2016   The Minamata Convention on Mercury is endorsed by the Cabinet.

The Kumamoto prefectural government certifies an 80-something woman who lives outside Kumamoto as a Minamata disease patient. (The certified patients in Kumamoto Prefecture: 1,787 people)

March 2016   A woman in her eighties in Niigata is certified as a

Minamata disease patient. (The certified patients in Niigata Prefecture: 705 people.)

May 2016 The Kumamoto prefectural government certifies a 60-something woman and an 80-something man as Minamata disease patients. (The certified patients in Kumamoto Prefecture: 1,789 people)

September 2016 "The Forum: Sixty Years of Minamata Disease - Learn the lessons from industrial damage and work toward a sustainable society" is held by Chulalongkorn University, Kumamoto Gakuen University and others in Bangkok, Thailand.

February 2017 The group of fetal Minamata disease patients, named Wakakatta-Kanja-no-Kai: the association of the patients who were once young, holds a music concert by the singer, Sayuri Ishikawa.

May 2017 The film director Kazuo Hara makes a statement: "Even though they have a human figure, their inside is not human any more." People say in protest that his words hurt patients, then he makes an apology.

The plaintiff, F, who won the Kansai case at the Supreme Court, files a suit to ask for confirming the status based on the compensatory arrangement. The case is decided in favor of the plaintiff by the Osaka District Court.

August 2017 The Minamata Convention on Mercury enters into force.

September 2017 In a reversal of a lower court ruling, the lawsuit for disability compensation for Minamata disease at the Supreme Court is decided against the plaintiff, Kawakami.

September 2017 A fetal Minamata disease patient, Shinobu Sakamoto attends the first meeting of the Conference of the Parties to the Minamata Convention on Mercury (COP1) as an NGO member and reports that Minamata disease has not ended.

November 2017   At the lawsuit demanding the mandamus action for the certification as Niigata Minamata disease patients, the Tokyo High Court orders Niigata Prefecture to certify all the nine plaintiffs as Minamata disease patients.

December 2017   The Niigata city government certifies nine people on the 15th and makes an apology. (The certified patients in Niigata Prefecture: 714 people)
The Supreme Court ruling in the lawsuit against the national, Kumamoto, and Kagoshima governments is against the plaintiff, who not only claims that Minamata disease is not a case of food intoxication under the Food Sanitation Law but also demands an investigation into Minamata disease.

January 2018   The Niigata Prefectural government closes the operation for the certification based on the Special Measures Law of Minamata disease. (1,829 people for the lump-sum payment, 140 people only for a Minamata Disease Victim's Notebook)

March 2018   Ruling in the third lawsuit on Niigata Minamata disease at the Tokyo High Court is against the plaintiff, and it is appealed to the Supreme Court in April.

March 2018   In a reversal of a lower court decision, the ruling in the case to ask for the status based on the compensatory arrangement at the Osaka High Court is against the plaintiff, F, who won the Kansai case at the Supreme Court before. He appeals to the Supreme Court in April.

May 2018   The president of Chisso, Goto, says at the memorial ceremony, "We offered a possible rescue based on the Special Measures. I believe everything is finished." On the 18th, due to widespread criticism, he withdraws the remark.

December 2018   The 75th issue of Tamashii Utsure, the Hongan-no-Kai newsletter, comes to its last publication.

The plaintiff, F, who won the case of the Kansai lawsuit at the Supreme Court, files a suit to ask for the status based on the compensatory arrangement at the Supreme Court. The ruling is against the plaintiff.

Students learning from Minamata (Photographed in 2014)

Minamata City Events

Month/Time of year		Events	Venues
1	Mid-Jan.	Kumanichi Santaro Relay Road Race	Minamata, Ashikita, Tsunagi
	Late Jan. to Feb.	Minamata Ashikita Ebi (Shrimp) Iro Iro Fair	the Minamata and Ashikita area
2	Early Feb. to mid-Feb.	Kugino Boar Meat Hot Pot Marathon	around Airinkan
	Early Feb. to mid-Feb.	Minamata Ashikita Japanese Drum Festival	Minamata City Cultural Hall
	Early Feb. to late Mar.	Minamata Spring Cuisine Festival	restaurants in Minamata City, the Yunoko and Yunotsuru hot spring inns
3	Early Mar.	Citizens' Relay Road Race	the Ecopark Minamata Athletic Stadium
		Performing Arts Festival in Orange-kan	Orange-kan
	Late Mar.	Saratama-chan Onion Festival	around Ecopark Minamata
	Late Mar.	Yunoko Onsen Sakura Festival	Yunoko
4	Mid-Apr.	Minamata Spring Festival	the Minamata shopping avenue
	Late Apr.	Shichijushichiya Celebration of Offering Minamata-Tea to the Gods	JA Minamata Tea Processing Center
	Late Apr.	Festival marking the start of boating season at Samukawa Suigen-tei	the Samukawa riverhead
	Apr. to early May	Trail of Minamata Umakamon (goody) in Spring	restaurants in Minamata City
5	1	Minamata Disease Victims Memorial Service	Otomezuka
	1	Minamata Disease Sufferers' Memorial Ceremony	Shinsui-ryokuchi at Ecopark Minamata
	Early to late May	Minamata Rose Festa in Spring	Rose Garden in Ecopark
	Mid-May	Tanada-no-Akari (lights on rice terrace)	the rice terrace in Samukawa, Kugino
	Late May	Minamata Rose Marathon	Eco-park Minamata
	Late May	Minamata Products Exhibition/ The Minamata Port Festival	Eco-park Minamata, the Minamata Port
	Late May	Minamata Food Festival	Furusato Hiroba at Ecopark Minamata
	Late May	Moyaikan Culture Festival in Spring	Moyaikan
	Late May to late June	Minamata Shirasu-don (whitebaits on rice) Fair	restaurants in Minamata City
6	One day in June	Nakaoyama Plum Picking	the Nakaoyama Plum Garden
	Early June	Yamabiko Music Festival	Kugino Eelementary School Gymnasium
	Mid-June	Moyai Music Festival	Minamata City Cultural Hall
7	Late July	Renryusai Festival	downtown Minamata
	Late July	Gion-san Festival	Marushima Shrine

8	Early Aug.	Serifune Competition (a traditional canoe race)	the mouth of the Minamata River
	Early Aug.	Saturday Night Market	downtown Minamata
	Mid-Aug.	Sakuragaoka Kannon Festival	Sakuragaoka Kannon-do
	Mid-Aug.	Shoro-nagashi (foliating lanterns for the spirits of the dead)	the levee of the Minamata River
	Mid-Aug.	Yunotsuru Onsen Summer Festival: the Bell-ringing Cricket Festival	the multi-purpose plaza of Yunotsuru Hot Spring
9	Early Sep. to Mid-Sep.	Women's Festival of Cutlassfish Fishing	Yunoko Hot Spring
	Early Sep.	Minamata Fireworks Festival	Yunoko Hot Spring
	Early Sep.	Minamata Yosakoi Festival	three venues in Minamata City
	Late Sep.	Hinomatsuri (Festival of Fire)	the bank harmonized with water at Ecopark Minamata
10	Early Oct.	Minamata Sports Festival/ Citizens' Sport Meet	Eco-park Minamata
	Early Oct.	Nakaoyama Cosmos Festival	Nakaoyama Park
	Mid-Oct.	Fukuro Fureai Festival	Fukuro-tenmangu Shrine
	Mid-Oct.	Sanno Jinja Festival	Sanno Shrine
	Mid-Oct.	Ozono Kojin-san Autumn Festival	Kojin-jinja Shrine
	Mid-Oct.	Gongen-san Festival	Gongengu Shrine
	Mid-Oct.	Kugino Sumiyoshi Jinja Autumn Festival/ the Sumo ring entering ceremony for babies	Kugino Sumiyoshi Shrine
	Mid-Oct.	The Eve of Ebisu Festival	the parking lot next to Ebisu Shrine
	Late Oct.	Mina Tosho Festival	the Municipal Library and the City Hall
	Late Oct. to late Nov.	Minamata Shirasu-don (whitebaits on rice) Fair in Autumn	restaurants in Minamata City
11	Early Nov.	Soho Fudezuka Festival	Soho-fudezuka-mae
	Early Nov.	Fukuro Tenmangu Autumn Festival/ the Sumo ring entering ceremony for babies	Fukuro-tenmangu Shrine
	Early Nov.	Citizens' Cultural Festival	Minamata City Cultural Hall, Moyai-kan
	Early Nov.	Yunotsuru-onsen Festival of Autumn Leaves	Yunotsuru Hot Spring
	Early Nov.	Minamata Japanese Black Tea Fair	
	Early Nov. to late Nov.	Minamata Rose Festa in Autumn	Eco-park Minamata
	Mid-Nov.	Minamata Industrial Complexes Festival	the Minamata Industrial Complexes
	Late Nov.	Yunotsuru Festival of Autumn Leaves	Yunotsuru
	Late Nov.	Hakusai (Chinese cabbage) Festival	the field near the bus stop, Nogawa
	Late Nov.	Asahi-machi Shopping Avenue Autumn Festival	the Asahi-machi shopping avenue
12	Early Dec.	Minamata Sports Festival/ New Sports Meet for Exchange	Minamata City General Gymnasium
	Early Dec.	Eco-park Minamata Illumination	Eco-park Minamata

	Early Dec.	Fureai Marathon	Eco-park Minamata
Monthly	Every fourth Saturday	Minamata Fresh Market	Michi-no-Eki Minamata
Monthly	Every second Saturday	Minamata Fishermen's Market	the Marushima New Port

https://www.go-minamata.jp/category/nature.html
The above event information of Minamata City is listed in the Minamata Newsletter on March 15.

Current Situation in Minamata
～ "Lesson from the Failure" ～

Over half a century has passed since the official recognition of Minamata disease. The sufferers who consumed mercury-polluted fish and shellfish have raised their voices and are fighting against national and prefectural governments as well as Chisso even after 64 years.

Measures to establish a social consensus by facing the real lifestyles of the sufferers, grasping the whole image of mercury pollution based on the process of "democracy", preventing expansion of the damage, discussing compensation, providing support for their daily lives, and eliminating discrimination and prejudice have not been done sufficiently. And because of repeated improvisation by national and Kumamoto prefectural governments, sufferers have been separated. With the Chisso Company being supported by the government, Minamata had been controlled economically and mentally by Chisso. Under such circumstances, the damage of Minamata disease was expanded. However, the Minamata disease incident happened in the poor fishing village far from "the center" (Tokyo), and as a result was trivialized by national and Kumamoto governments and the dignity of each sufferer was ignored.

What we need to do now is to face both the history and the present situation of the Minamata disease incident which have been opened up by the sufferers once again. "Learning from the failure of history and reviving the message from the Minamata disease incident into the future", which is the dying wish of late Dr. Harada who advocated the Minamata Study, is the minimum responsibility for us to do for future generations. "Reviving the message from the failure" means every citizen should pay attention to the issue of the area and act towards its solution regardless of fragmented expertise. Furthermore, act beyond the framework of amateurs and professionals. At the same time, when the social and economic system is destroyed at times such as in the Fukushima nuclear power plant accident after the Great East Japan earthquake, a flexible approach to development or recovery of the area is required.

In Minamata we have affluent nature - rivers which connect the mountains to the sea. Local government, citizens, and companies all together have sought safe and secure manufacturing and ways of life considering protection of the environment since the "Environmental Model City Declaration" was made in 1992.

The project of the "Whole Village Life Museum" (currently four areas: Kagumeishi, Okawa, Koshikoba and Kugino) considers the valuable connections between people and nature based on lifestyle and culture. In the Kugino area, utilizing former Kugino Station on the old JR Yamano Line, persistent activities such as developing forests for water sources, lighting at paddy fields, gatherings which feature home-style cuisine, and boar meat hot pot marathons take place under the theme of 'village development based on ecology' (natural features, circulation, and independence).

In addition, citizens, governments, and businesses have been united in continuous efforts on waste, symbolized by the "20-category separation of garbage (as of April 2018)," which has attracted attention from all over the country. In addition to reducing landfill garbage and effectively using resources (recycling), this developing initiative aims to deepen the bonds between people in the community and create a new lifestyle that is conscious of the shift from recycling to reuse.

The city of Minamata won the title of Japan's "Environmental Capital" at "the 10th Japan Environmental Capital Contest" as the first awardee in the nation in March 2011. The city is expected to lead "the National Eco-City Contest Network", established as a strategic network for local governments, NGOs, research institutes, and various other groups to join and work in December 2012 (it consisted of 17 local governments, 16 NGOs, 6 researchers of 6 research institutes as of April 2014).

Along with enforcement of the Special Measures Act, which aimed to relieve Minamata disease sufferers and solve the Minamata disease issue in 2009, the national government began financial support to the Minamata disease-affected areas. In 2011, the "Minamata Environmental Town Development Research Group" organized by consultants and persons with relevant knowledge and experience in Tokyo and wrote the report. With a change of mayor, the priority of the policy of the Minamata City government has been reviewed and environmental concerns have been put on the back burner. It seems the citizen-initiated town development is taking a step back. Now is the time to revive "the message from the failure".

Inauguration of the National Eco-City Contest Network（Photographed in 2012）

Onsite Research Center for Minamata Studies, Kumamoto Gakuen University
Map1 F-5

The Onsite Research Center for Minamata Studies was established in August 2005 as a research hub to connect the field of Minamata with the Open Research Center for Minamata Studies of Kumamoto Gakuen University (in Oe, Chuo-ku, Kumamoto City) which was inaugurated in April of the same year.

"Minamata Studies" is an academic discipline which transcends academic frameworks (i.e. interdisciplinary), which is open to everyone, and is beyond the border of "amateur" and "professional". It is also a field of study based on an abundance of facts, which redefines the lifestyle of each one of us. The results and achievements of our study are returned to the community and disseminated to the world.

Our activities are based on the three pillar projects; Comprehensive study which is to clarify the entire picture of damage by mercury pollution and to discover solutions to the problems; Social experimental study which is to assess local rehabilitation in the communities after conquering environmental damage and to build democratic consensus; Being an international information center by accumulating and organizing a database through archives and to disseminate materials related to Minamata studies.

From 2009, our biography has been shown on our homepage as "Minamata Study Database". What is highly noteworthy is the ex-library document from the labor union of Chisso Corporation. The copies of documents by labor union members who took the witness stand on the side of patients, and accused the corporation of its crimes in the first Minamata disease litigation, which were kept by former labor union members are available at the Onsite Research Center for Minamata Studies. The Minamata Archives are also shown on our homepage where people can study Minamata by images and photos.

To open the research center to the community, we provide extension courses to citizens (tuition free) every fall. In addition, the

Center regularly offers counseling services on healthcare and welfare.

Our Center was chosen as an institute supported by the Open Research Center Project for the Private Universities (2005~2009), (2010~2014) and (2015~2019) with a matching subsidy from the Ministry of Education, Culture, Sports, Science and Technology. The Center functions as the research base on Minamata studies for faculty members and graduate students of Kumamoto Gakuen University. In addition, we hope that the Center is openly used by researchers in Japan and overseas, as well as by community members as a common space to enjoy mutual communication and exchange.

Open: Tuesday ~ Friday
Opening hours: 10:00~16:00
Parking space: 10 cars
Address: 2-7-13 Hama-machi, Minamata, Kumamoto, 867-0065
Phone: 0966-63-5030
Fax: 0966-83-8883
E-mail: m-genchi@kumagaku.ac.jp